POWER PLAYS

Homeland Security Series

Series Editors:
Tom Payne, University of Southern Mississippi, USA
Tom Lansford, University of Southern Mississippi, USA

This series seeks to provide a body of case studies to explore the growing importance and prominence of homeland security to national defence policy and to examine the development of homeland security within the broader context of national defence policy in the United States and other major developed states. This series will identify and analyze the major threats that are particular to homeland security, as well as those that affect broader national security interests. Comparative studies will be used to elucidate the major similarities and differences in how states approach homeland security and works which advocate new or non-traditional approaches to homeland security. The series aims to integrate information from scholars and practitioners to provide works which will influence the policy debate and examine the ramifications of policy.

Also in the series

Calculated Risks
Highly Radioactive Waste and Homeland Security
Kenneth A. Rogers and Marvin G. Kingsley
ISBN 978-0-7546-7133-6

The State and Terrorism
National Security and the Mobilization of Power
Joseph H. Campos II
ISBN 978-0-7546-7192-3

Comparative Legal Approaches to Homeland Security and Anti-Terrorism
James Beckman
ISBN 978-0-7546-4651-8

To Protect and Defend
US Homeland Security Policy
Tom Lansford, Robert J. Pauly, Jr and Jack Covarrubias
ISBN 978-0-7546-4505-4

Power Plays
Enriched Uranium and Homeland Security

CHRISTOPHER HUBBARD
Curtin University of Technology, Australia

Routledge
Taylor & Francis Group
LONDON AND NEW YORK

First published 2010 by Ashgate Publishing

Published 2016 by Routledge
2 Park Square, Milton Park, Abingdon, Oxon OX14 4RN
711 Third Avenue, New York, NY 10017, USA

Routledge is an imprint of the Taylor & Francis Group, an informa business

British Library Cataloguing in Publication Data
Hubbard, Christopher, 1953-
Power plays : enriched uranium and homeland security. --
(Homeland security)
1. Nuclear industry--Security measures. 2. Uranium
enrichment--Safety measures. 3. Uranium enrichment--
Government policy. 4. National security. 5. Nuclear
nonproliferation--International cooperation.
I. Title II. Series
363.1'79956-dc22

Library of Congress Cataloging-in-Publication Data
Hubbard, Christopher, 1953-
Power plays : enriched uranium and homeland security / by Christopher Hubbard.
p. cm. -- (Homeland security)
Includes index.
ISBN 978-0-7546-7693-5 (hardback) -- ISBN 978-0-7546-9392-5 (ebook)
1. Highly enriched uranium--Government policy--United States. 2. Highly enriched uranium--Safety measures. 3. Uranium enrichment--Government policy--United States. 4. Uranium enrichment--Safety measures. 5. National security--United States. 6. Nuclear nonproliferation--Government policy--United States. 7. Nuclear terrorism--United States--Prevention. I. Title.
TK9152.165.H835 2010
363.17'995610973--dc22

200904713

ISBN 13: 978-0-75467-693-5 (hbk)

Contents

Preface *vii*
List of Abbreviations *xi*

1 Introduction: Enriched Uranium and Homeland Security 1

2 Securing the Homeland, Securing the World 19

3 The International Institutional Framework 39

4 Assessing Security Threats Inherent in Nuclear Fuels 59

5 The Historical Record: Failure and Success 79

6 Strategic Security and Enriched Uranium 97

7 Nuclear Terrorism and Enriched Uranium 115

8 Old and New Strategies for Non-proliferation Progress 133

9 The Future: Alternative Policy Pathways for the Twenty-first Century 149

10 Conclusions: Global Threats and Opportunities 165

Appendix A: Treaty on the Non-proliferation of Nuclear Weapons *181*

Index *187*

For Paul, Pamela, Richard and Jenny

Preface

In a letter from its Co-Chairman in the opening pages of the Annual Report of the think tank the Nuclear Threat Initiative for 2007, entitled *Vision and Action for a Safer World*, former US Senator Sam Nunn and philanthropist Ted Turner made the following startling claim:

> In October of 2001, top US government officials received a highly classified intelligence report, warning that terrorists had acquired a 10-kiloton nuclear bomb and planned to smuggle it into New York City, where it could-if detonated-destroy much of lower Manhattan and kill hundreds of thousands of people. This intelligence report was later judged to be false. But it was never judged to be implausible. (Nuclear Threat Initiative 2007a, 2)

The possibility of an unannounced and devastating nuclear attack on a crowded city by a terrorist group, or indeed the armed forces of a nation state, is the nightmare from which the world may not awaken. In 2010, the odds of such a tragedy happening are growing ever shorter. Like it or not, the know-how necessary to mount an attack of this kind already exists in respect of a number of vulnerable technologies and practices, such as uranium enrichment, the transportation and storage of enriched uranium and plutonium, and the widely available engineering and metallurgy needed to construct crude nuclear weapons.

Early in 2008, former United States Secretaries of State George Schulz, William Perry and Henry Kissinger, together with Senator Nunn, called in *The Wall Street Journal* for urgent and concerted global action to deny operational nuclear weapons to terrorists, and ultimately end their nuclear threat to the entire world. Among a range of other initiatives, these distinguished statesmen called for dramatically higher standards to be applied to the security of nuclear weapons and materials in the more than 40 countries throughout the world which hold them. In addition, they called for an international program to manage the weapons proliferation risks accompanying the nuclear fuel cycle in an era of growing interest in nuclear energy.

While the United States and Russia are responsible for some 95 per cent of the world's nuclear warheads, and occupy central roles in the evolution of global nuclear arms control and disarmament initiatives, as well as non-proliferation diplomatic architectures and nuclear technologies, they alone cannot succeed in stemming the tide of nuclear weapons diffusion. The discussions and analyses which follow will help to explain why this is so, while also helping to shed some more, much needed, light on the nature and scope of what still remains to be done.

The central aim of this book is to offer a comprehensive and concise exploration of the effects on national/state security policy and discourse of the production, storage, transportation, safeguarding, export and use of enriched uranium and – by extension – plutonium. Its examination will be developed with the needs of the informed generalist in mind, and thus will avoid unnecessarily complex discussion of highly technical matters, as well as the professional nuclear-related jargon which often impedes popular understanding. Nevertheless, this analysis examines a wide range of interlinked geopolitical, security and technical issues and challenges for national and global governance in the dual fields of nuclear power generation and the proliferation of nuclear weapons materials and technologies.

Within this rubric, the central question for resolution is how best to strike the most effective balance between alternative and complementary policy approaches necessary to meet the most important security issues inherent in the physical properties of uranium and plutonium.

The theoretical background for the book's analysis combines the claims and insights of the English School's 'International Society' paradigm, led by Bull, with the broadening and deepening security agenda of Buzan et al.'s Copenhagen School of studies in critical security. In addition, the analytical strategy of the book is to use comparative analytical techniques to highlight both the universal and contingent security elements relevant to enriched uranium and plutonium, as well as the ways in which relevant threat perceptions are generated.

The state-centric foci of analysis are developed as three sets of nuclear-interested countries: established nuclear nations such as the five declared nuclear weapon states and others (for example, South Korea, Sweden, Canada, Japan and Spain) with mature nuclear power industries; developing states with nuclear weapons (such as North Korea, India and Pakistan); and aspirant or clandestine nuclear nations (such as Iran and Israel).

A fourth – arguably the most critically important – focus of all is the range of relevant non-state agents (for example, terrorist groups, the nuclear scientific community and non-governmental organizations). Finally, it can be noted here that structural, institutional, historical and human agency factors will each find a place in the book's analysis.

Turning to uranium and its enrichment, most of the approximately 470 commercial nuclear power reactors currently operating or under construction globally need fuel which is "enriched" in U-235, the only naturally occurring isotope of uranium, present in usable quantities, which is capable of producing a nuclear chain reaction (or "fission"). The need for enrichment stems from the fact that around 99.2 per cent of natural uranium consists of its isotope U-238, which does not contribute directly to the fission process.

As a result, the naturally occurring percentage of fissile U-235 found in mined uranium (around 0.72 per cent) must in most cases be boosted, or enriched, in order to produce a sustained and controlled fission reaction releasing energy in the form of heat. For use in power reactors, the enrichment of U-235 is normally in the order of 3–5 per cent, and this modest level of enrichment will not produce uranium suitable for use in nuclear weapons. However, further U-235 enrichment to the 90 per cent level, or greater, which is necessary for weapons use can be reached simply by continuing the enrichment process, typically using large gas centrifuge cascades.

It is here that the rationale of the book is most clearly evident: uranium left in the ground, used in its natural form or slightly enriched in its fissile isotope U-235 presents no insurmountable military or terrorist security risks when it, and its waste products, are employed and handled safely and securely in the production of heat (and thus electrical power) in a nuclear power generation plant.

However, when uranium is enriched beyond the level needed for power generation, its legitimate applications – which in effect include any worthwhile application beyond its use as the explosive "fuel" for nuclear weapons – are highly specialized and few in number; one important application is as a fissioning "target" in the manufacture of medical radio-isotopes. The same quality of highly restrictive legitimate use is also true of the weapons-grade plutonium (mostly consisting of its isotope Pu-239) which is produced as a residual or by-product of the uranium fission process in nearly all power reactors.

There is, then, a vital need for policy makers and their advisers to understand the homeland security implications for any country holding, manufacturing, transporting, securing, importing or exporting enriched uranium or plutonium, either for itself or on behalf of other countries, or that believes its national security to be compromised by similar activities in neighboring states. Those states which export natural uranium for use in power generation, such as Australia and Canada, must also calculate the downstream security implications of their activities in terms of the clandestine diversion of the uranium they supply away from power generation and into nuclear weapons programs. The continuing and accelerating trend towards the proliferation of nuclear weapons, their materials and technologies, looms ever larger in the panoply of existential risk assessments for many states. It is a trend which, moreover, can only be successfully resisted by coordinated global action denying to proliferators the fissionable nuclear material without which a completed nuclear weapon is merely an exercise in engineering, electronics and metallurgy.

The nature and dangers inherent in the open or "once through" nuclear fuel cycle, from uranium mining to the short term storage and long term geological sequestration of irradiated highly enriched uranium, plutonium and other high radiation level nuclear wastes, are central to a wide range of serious security concerns at both the global and US homeland security levels. Furthermore, the domestic and international aspects of ensuring effective nuclear security are inextricably and interdependently linked. The need for policy attention directed towards addressing the security issues created by nuclear materials will only strengthen over time, as renewed interest in nuclear power generation translates into the construction of historically large numbers of new nuclear power plants around the world – including those now planned or building in the United States, China, India and several other states.

In parallel with renewed nuclear interest generally, an active renaissance is now underway in the development of the processes of uranium enrichment itself, in order to harness new technologies emerging in response to the anticipated rise in worldwide demand and production of nuclear fuels. In short, the use of uranium around the world for power generation will not dissipate any time soon, and is likely to grow strongly during the coming decades. Thus, the issues addressed here will continue to be actively and increasingly relevant to global security.

As noted above, it is to be hoped that this book will contribute in a modest way to popular levels of understanding about one of the most serious of the security implications inherent in the present rapid growth in nuclear power generation capacities across the world. International proposals to limit the accelerating global proliferation of nuclear weapons technologies and capacities now turn emphatically on denying both proliferators and terrorists access to uranium usable in weapons development programs or improvised nuclear devices, and to operational weapons themselves. If successful, they will do so, at least in the first instance, largely through the development of new strategies aimed at denying potential proliferators, and their terrorist clients, access to the capacity to illicitly enrich uranium beyond the reach of international oversight. One example is embodied in recent proposals (such as the US Global Nuclear Energy Partnership) to develop international enrichment centers providing secure enrichment services to states while employing new, proliferation-resistant enrichment technologies.

It is the book's purpose to examine the chances for success of proposals such as this, as well as the costs of failing to account safely for the latent dangers to homeland security posed by the diversion beyond the nuclear fuel cycle, and subsequent misuse, of enriched uranium and plutonium.

Christopher Hubbard

List of Abbreviations

BRIC	Brazil, Russia, India, China Group of Countries
CANDU	Canadian Deuterium-Uranium Reactor Technology
CTBT	Comprehensive Test Ban Treaty
CTR	Nunn-Lugar Cooperative Threat Reduction Program
CTWG	US-Russia Working Group on Counter-terrorism and Strategic Security
DOD	United States Department of Defense
DOE	United States Department of Energy
DHS	United States Department of Homeland Security
DPRK	Democratic People's Republic of Korea
EMP	Electro-magnetic Pulse
FATA	Federally Administered Tribal Area
FMCT	Fissile Material Cut-off Treaty
G8-GP	G8 Global Partnership Against the Spread of Weapons and Materials of Mass Destruction
GIF	Generation IV International Forum
GNEP	Global Nuclear Energy Partnership
GTRI	Global Threat Reduction Initiative
GW (e)	Gigawatts-electric
HCOC	Hague Code of Conduct Against Ballistic Missile Proliferation
HE	High Explosive
HEU	Highly Enriched Uranium
IAEA	International Atomic Energy Agency
ICBM	Inter-continental Ballistic Missile
IND	Improvised Nuclear Device
IPFM	International Panel on Fissile Materials
IRBM	Intermediate Range Ballistic Missile
ITDB	Illicit Trafficking Data Base
kt	Kiloton of TNT Explosive Equivalent
LEU	Low Enriched Uranium
LOSC	Convention on the Law of the Sea
MNA	Multilateral Nuclear Approach
MOX	Mixed Oxide Fuel
Mt	Megaton
MTCR	Missile Technology Control Regime
MW (e)	Megawatts electric
MW (th)	Megawatts thermal

NATO	North Atlantic Treaty Organization
NNA	United States Nuclear Non-proliferation Act, 1978
NNSA	National Nuclear Security Agency
NNWS	Non Nuclear Weapon State
NPT	Nuclear Non-proliferation Treaty
NSG	Nuclear Suppliers Group
NTI	Nuclear Threat Initiative
NWS	Nuclear Weapon State
OMB	Office of Management and Budget
PNE	Peaceful Nuclear Explosion
Pu-239	Plutonium Isotope 239
PSI	Proliferation Security Initiative
SORT	Strategic Offensive Reduction ("Moscow") Treaty
START	Strategic Arms Reduction Treaty
TNT	Trinitrotoluene Explosive
U-235	Uranium Isotope 235
U-238	Uranium Isotope 238
UNAEC	United Nations Atomic Energy Commission
UNMOVIC	United Nations Monitoring, Inspection and Verification Commission
UNSC	United Nations Security Council
UNSCR	United Nations Security Council Resolution
WMD	Weapon of Mass Destruction
WNA	World Nuclear Association

Chapter 1
Introduction: Enriched Uranium and Homeland Security

First Steps

As noted in the Preface, the central aim of this book is to offer a comprehensive exploration of the effects on homeland security – a term whose meaning will be examined below – of the production, safeguarding, transportation, import, export, and use of enriched uranium, and – by extension – plutonium. Its primary motivation is the increasingly pressing need for coordinated action on a global scale to develop effective strategies denying to state proliferators and terrorist groups the capacity to pursue their nuclear explosives goal. When national governments or international terrorist organizations seek access to the ultimate weapon of coercion, the world's people must insist on the greatest possible range of responses to a trend with the ability, if left uncontrolled, to permanently destroy human society and, indeed, life on Earth.

In undertaking this complex and important task, the discussions which follow look closely at a wide range of interdependent, geo-strategic, political, technical and security-related issues and challenges for both national and global governance in the dual fields of nuclear power generation and the proliferation and deployment of nuclear weapons materials and technologies. The central conundrum inherent in such a quest is not whether the two elements of applied nuclear physics can co-exist in a harmonized way which leads, eventually, to the eradication of nuclear weapons. Rather, it lies in the question how the greatest danger of all, which is complacency among governments and peoples over burgeoning nuclear risks, can be overcome in ways which instill a viscerally real appreciation of the nature of the threats posed by the further diffusion of nuclear weapons technologies and materials.

However, in order to engage fully with the complexity of present and future security implications arising in this context, it is essential to draw the broad outlines of current and anticipated global and regional geo-political and geo-strategic settings and trends. While necessarily generalized and sweeping in nature, such an exercise helps to establish the parameters and limiting factors present in the nuclear security challenges the world is now facing. Furthermore, it will be useful at the outset to incorporate into this discussion the character and relevance of the notion of "Homeland Security" *per se*, both for the United States itself and for other states seeking nuclear and other security solutions for their own people's future.

The Homeland Security Institute, a research and analysis organization authorized by the US Homeland Security Act of 2002, has summarized the nature of homeland security for the United States and, in doing so, has encapsulated the central purpose and accompanying duties of all states. The purpose of the American state and nation, as found in its Constitution's Preamble, is to:

> ... provide for the common defense, promote the general welfare, and to secure the blessings of liberty for ourselves and our posterity.

United States institutions and organizations at the federal, state, local and private levels all undertake these duties, while governmental functions are grouped around three broad purposes:

- *Homeland Security* – protecting the common good domestically against attack
- *General Welfare* – those actions that advance the common good domestically, and
- *National Security* – those actions which both protect and advance the common good internationally.

Within this rubric, the Institute defines Homeland Security within three sub-sets:

- *Homeland Security* itself: the deterrence, prevention, and preemption of, and defense against, aggression targeted at US territory, sovereignty, population, and infrastructure as well as the management of the consequences of such aggression and other domestic emergencies.
  - Homeland Defense, a subset of Homeland Security, defined as the deterrence, prevention and preemption of, and defense against, direct attacks aimed at US territory, population, and infrastructure.
  - Civil Support, also a subset of Homeland Security, and separate from Homeland Defense. It is defined as Department of Defense (DOD) support to civilian authorities for natural and man-made domestic emergencies, civil disturbances, and designated law enforcement efforts.

As the Institute for Homeland Security points out,[1] national focus on each aspect of US homeland security has changed considerably since Independence, waxing and waning as a pressing need for homeland defense has in turn been supplanted by greater emphasis on enhancing the general welfare through trade and other economic activity. During the nineteenth thirties, the need to deal with the effects

1 Larsen, R. and McIntyre, D., ANSER Institute for Homeland Security, 2008. Available online: http://www.homelandsecurity.org/bulletin/overview.htm. Accessed November 16, 2008.

of the Great Depression for an America which felt itself safe behind two oceans eventually led to an about-face as the Second World War demanded overwhelming emphasis on national security. In 2010, with the Cold War a fading memory and asymmetrical attack by terrorist groups and unpredictable states the central security focus, the pendulum has swung once more towards homeland security and civil support.

This security taxonomy bears the hallmarks of the fundamental security strategy of any nation state worthy of the name, and of the aspirations and commitments of its government and people. No plural liberal democracy will unilaterally divest itself of a careful and durable focus on the goals of the United States' homeland security, homeland defense or civil support paradigms. Although emphases will vary around the international system (and sometimes considerably in the case, for example, of policies towards deterrence and preemption strategies) nevertheless, it is appropriate for states to use US practice in planning their own domestic security as both a benchmark and sounding board for parallel activities in their territories and jurisdictions.

Global Security – Evolving Change

The international system is now changing at an accelerating rate, a process which poses an increasingly wide range of security challenges for the nation states, a diverse group of non-state agents, and for individual people. Recently described as the "global jigsaw,"[2] the complexities increasingly apparent within the global security equation considered on its broadest scale are belied somewhat by the appearance of a set of countervailing movements – undeniable trends which continue to strengthen as they mark and clear the world's path to the future. This brief summary of global strategic change and continuity is offered here merely as an introductory note.

First, the US will remain as the unchallenged global hegemon for perhaps the next twenty years or so, while gradually surrendering its relative influence and agency to newly ascendant power centers such as Brazil, Russia, India and China (the "BRIC" states). The recent and spectacular economic success of these states, and especially of India and China, promises to bring about a re-balancing of multi-polar global power as wealth creation inevitably translates into growing military and other coercive capacities and ambitions.

Second, globalizing processes continue simultaneously to intensify the economic integration of the Westphalian world of nation states while fragmenting it politically. In the process, the very coherence and durability of the states are coming under growing attack from transnational and non-state actors. At the same

2 Lyon, R. and Leah, C., *Global jigsaw: ASPI's strategic assessment 2008*, Australian Strategic Policy Institute, October, 2008. Available online: http://www.aspi.org.au/publications/publicationlist.aspx?pubtype=5. At pp. 3, 4. Accessed December 11, 2008.

time, many traditionally understood and accepted multilateral architectures and organizations of global governance, cooperation and coordination are becoming ever less relevant as the international system itself transforms.

Third, the possibility of general war between the states using the full range of modern weapons, including weapons of mass destruction, remains real, although less likely over time. Furthermore, it is increasingly probable that such a conflict would be the result of asymmetrical warfare, initiated either by terrorist organizations or uncontrolled military regimes in nation states. Concurrent with these trends, nuclear weapons proliferation appears to pose as intractable a threat as ever while it evolves new dimensions, such as the acquisition and use by small non-state groups of inadequately secured weapons technologies and fuels.

Fourth, absolute world population growth, accompanied by large disparities in demographic trends between the developed and the developing worlds, hold out the risk of economic disruption, internal conflict and even state collapse for some vulnerable states such as those, for example, in Sub-Saharan Africa. New circumstances of these kinds offer the possibility of safe haven for terrorists seeking to attack the West. Furthermore, contracting and aging populations in countries such as Japan and many European countries, mean that these wealthy states will find it increasingly difficult to grow, or even maintain, the strength and resilience of their economies and their capacity to resist nuclear threats.

Finally, a transnational agenda has appeared as a result of global economic success, placing ever-growing strains on several vital classes of resources, chief among them food, water and energy supplies. The World Bank has suggested that demand for food is set to rise by some 50 per cent by 2030 as a direct consequence of rising absolute population levels, growing affluence in some regions, and the appearance of new middle classes able to afford – and thus compete for – more and better food and water supplies.[3]

A second element of the transnational agenda is the extended transition from carbon-based energy economies to those driven by alternative non-carbon energy sources, including nuclear and diverse renewable power sources. Climate changes as a result of atmospheric warming, as carbon dioxide and other "greenhouse" gas emission levels continue their upward spiral, must eventually result in widespread redistribution and disruption of food producing regions.

Finally, the diffusion of new and innovative technologies will be both a blessing and a curse on the global scale. The ready availability of digitally manipulated information from the Internet, data which in earlier times could be closely held and safeguarded, is increasing evading the capacities of states to control its distribution. Terrorist operations planned and executed by virtual and ephemeral groups across the globe, organized and linked in this way, will be extremely difficult to counter at the state level. By the same token, it is likely that governments will be less

3 National Intelligence Council, *Global trends 2025: A transformed world, Executive Summary*. Available online: http://www.dni.gov/nic/NIC_2025_project.html. Accessed December 11, 2008.

successful in coordinating global responses to terrorist threats when their security strategies and perceived national interests fail to align sufficiently closely.

In short, the world is facing a growing panoply of uncertainties and challenges in the context of deepening complexity and accelerating rates of transformation, whether in terms of terrorist threats, climate change, demographic transition, strategic re-balancing, nuclear weapons proliferation, resource sustainability or energy security. The next twenty-five years will be an era in which a root-and-branch reordering of the international system will become an unavoidable certainty forced upon a reluctant but fearful world. Equally inevitably, winners and losers will appear, either to seize the day or try to hold the line against a tsunami of challenge.

Enriched Uranium and Plutonium: Roles and Uses

Uranium, whether or not it is enriched in its fissile isotope U-235, and plutonium, are metallic chemical elements which lie at the heart of any discussion about nuclear affairs. This is true whether the focus is on the generation of electrical power using nuclear fission to produce heat, or on the development of nuclear weapons. In both cases, uranium and plutonium can be used as the necessary fissile material through their capacity, under specific conditions, to sustain a nuclear "chain reaction". It is notable, though, that both uranium (in its natural, non-irradiated form) and plutonium are only moderately radioactive.

A fission chain reaction, in either power generation or weapons applications, can be characterized as the cumulative absorption of relatively slow moving neutron particles (which have no electrical charge) by an atom, for example, of uranium-235, thereby splitting that atom and producing new neutrons. Those neutrons then bombard other fissile uranium atoms in the same way, creating more neutrons and forming a nuclear chain reaction. In the process, enormous quantities of thermal, electro-magnetic and kinetic energy are released. As an uncontrolled reaction, this takes the form of an extremely powerful explosion. When controlled and appropriately moderated in a reactor vessel, using a neutron-absorbing material such as graphite or heavy water, the abundant heat energy produced is used to created super-heated steam at high pressure, which in turn powers standard steam turbines linked to electrical generators, in order to generate electricity in the conventional manner.

At this stage, we can begin to see the nature of the nuclear security conundrum. The processes by which uranium and plutonium produce energy in nuclear weapons and electricity generating plants are essentially identical, although these two uses could hardly be more distinct in form and purpose; simultaneously, they serve both the warriors and the peacemakers. Similarly, there are significant and divergent implications for security policy planners when considering how best to counter the misuse of nuclear materials when used for weapons manufacture, as opposed to electricity generation. What is clear, however, is that in both cases severe and growing threats to homeland security continue to emerge for the United States, its Western allies, other nation states and, increasingly, a suite of newly nuclear-interested and relevant states.

The heart of the problem lies in the still abundant quantities of weapons-usable fissile material, in the form of both highly enriched uranium and plutonium, which are stockpiled around the world. The nuclear armed states continue to hold sufficient material for many thousands of nuclear weapons *beyond* their own stated needs, while large stocks of both highly enriched uranium and plutonium exist as fuel for over one hundred research reactors around the world, and as a stored by-product of uranium power reactors. As outlined by the 2007 Global Fissile Material Report of the International Panel on Fissile Materials, securing, consolidating and eliminating fissile material stocks worldwide are common imperatives in overlapping efforts to eliminate nuclear weapons in the countries where they exist, halt their spread to still more countries, and prevent terrorists from obtaining them.[4]

In order to make sense of the nature of the bilateral security challenge, we will take each case – power generation and weapons manufacture – in turn, beginning with nuclear power generation. It should be noted that this introductory overview serves in Chapter 1 only as a foundation for more comprehensive later analyses of the issues involved.

Both highly enriched uranium and plutonium are variously produced, stored, transported, safeguarded and utilized in many countries around the world, often in circumstances in which they are more or less vulnerable to diversion from legitimate use in power generation and scientific research towards illicit nuclear weapons programs. Such diversion can occur through the decisions and actions of the governments of states, or from the unilateral action of terrorist and other non-state groups and institutions; the Pakistan-based A.Q. Khan network, to be discussed later, is a prime example of the second case.

The Nuclear Fuel Cycle

The series of industrial processes by which enriched uranium and plutonium are used to produce energy is a unique setting for both efficient, low (total) carbon emissions power production processes, and the illicit diversion of the two primary types of nuclear fuels for use as nuclear explosives. In order to fully appreciate the significance of threats posed in these ways, a general comprehension of these processes is an essential first step.

4 International Panel on Fissile Materials, Global Fissile Material Report 2007, *Second report of the International Panel on Fissile Materials developing the technical basis for policy initiatives to secure and irreversibly reduce stocks of nuclear weapons and fissile materials*. Available online: http://www.fissilematerials.org. Accessed December 12, 2008. At p.2.

The World Nuclear Association, whose member commercial corporations produce around 85 per cent of the world's nuclear power, provides a concise summary of the nuclear fuel cycle.[5] Its major components can be summarized here. The cycle begins with the mining of uranium, a metallic element which is found throughout the earth's crust, using conventional mining techniques to extract it from economically viable uranium ore concentrations. Uranium is around 500 times as common as gold, roughly as common as tin, and is found in most rocks and soils, as well as in many rivers and all seawater. Around 85 per cent of the relatively low level of radioactivity associated with uranium in its natural state is the result of its radioactive decay processes. The milling process removes from the ore the residual but active radiation-emitting elements responsible for most of the radioactivity, leaving the processed uranium as only minimally radioactive. As uranium is only slightly radioactive in its natural form, it poses no significant difficulties within the initial mining process, whether in the course of standard excavation or alternative *in situ* leaching techniques.

Milling of the freshly mined uranium ore and simple chemical processing produce uranium oxide concentrate (known as "yellowcake") which usually comprises at least 80 per cent uranium. The yellowcake, being only slightly radioactive, can be readily and safely handled and transported using widely available mining and industrial equipment, without the need for special precautions. However, in this form it requires conversion into fuel suitable for use in a nuclear reactor, and must usually be enrichment in its fissile isotope U-235, as noted above. The residue, or tailings, from uranium milling pose some storage difficulties, since they contain a number of long-lived radioactive isotopes in low concentrations, and some heavy metals. Tailings must be kept isolated for long periods from their surrounding environment, usually in pits adjacent to mining operations.

For reactor types which use non-enriched uranium (such as those using the Canadian CANDU technology) the uranium oxide yellowcake, refined into uranium dioxide, can be immediately fabricated into reactor fuel. For most reactor types requiring enriched uranium fuel, however, the uranium dioxide is converted into uranium hexafluoride gas. Repeatedly stripped of its non-fissioning isotope U-238 using either a gaseous diffusion or gas centrifuge cascade process, the fissioning isotope U-235 is now present at a slightly higher concentration than that found in nature, typically around 3.5 to 5 per cent. In this form, it is reconverted into uranium dioxide to provide fuel for power generation in most nuclear reactors. The uranium dioxide is then sintered (or baked) at high temperatures to produce ceramic uranium fuel pellets. These are loaded into fuel rods, which in most cases are bundled together to form nuclear reactor fuel assemblies ready for loading into power reactors. As the fuel pellets are eventually depleted, or "burnt up," in their fissile uranium in reactor operations, some of the non-fissioning uranium-238 still present in the fuel is converted into plutonium, the main isotope of which

5 World Nuclear Association, *The nuclear fuel cycle.* Available online: http://www.world-nuclear.org/info/inf03.html. Accessed January 9, 2009.

(plutonium-239) is itself readily fissile. The result is that around one third of the energy generated in a typical nuclear reactor is produce by plutonium created in this way.

After a period of between one and two years, the bundled fuel is considered to be "spent," or depleted in its fissile isotopes of uranium and plutonium to such an extent that further use becomes impracticable. When removed from a reactor, the used fuel rods are emitting high levels of radiation and heat from a range of fission products (the result of splitting each large uranium atom into two smaller fragments). This nuclear waste material must be stored, often for several years, in ponds of ordinary water adjacent to the reactor to allow the spent fuel to cool while shielding its immediate environs against radiation. As the heat and radiation levels subside, the waste must at some stage either be prepared for semi-permanent storage and eventual geological sequestration, or for reprocessing into further nuclear fuel in a closed version of the nuclear fuel cycle.

Permanent disposal, or sequestration, of spent nuclear fuel and other high level wastes (such as the fission products noted above) has only in recent times begun to receive serious study and funding around the world. Several countries, including Sweden, Finland and, until 2009, the United States (at its Yucca Mountain, Nevada project) are developing permanent, high level waste disposal facilities with the aim of depositing wastes in large, stable geological formations at depths between 500 and 1,000 meters. If successful, nuclear wastes sequestrated in this way should remain fully isolated from the human environment in perpetuity. However, one important consideration in this approach is that an "open" fuel cycle such as this ignores the considerable potential for further extraction of energy through the reprocessing of spent fuel to recover its unused fissile uranium and plutonium fraction. Reprocessing, however, also produces significant quantities of high-level radioactive wastes.

A range of issues surround the complexities associated with reprocessing spent nuclear fuels, not the least of them being the potential for enhanced nuclear weapons non-proliferation outcomes. Another is the possibility of reducing the long-term radioactivity in high level nuclear wastes. Many of these and other considerations will be discussed in more detail at various points within this book, allowing the canvassing here of only the most important points. First, it can be noted that reprocessing spent uranium in fuel assemblies recovers around 25 per cent more energy from the original uranium than is obtainable by merely storing the spent fuel after its initial use. Furthermore, the one per cent of uranium-235 which remains unfissioned in spent uranium fuel, together with a similar percentage of plutonium created in the fission process, can be combined to form reprocessed "mixed oxide" or MOX fuel for use in suitable power reactors. The volume of high level waste produced by MOX fuel is only some one-fifth of that produced in an open fuel cycle, while the radiation levels present in the waste fall much more rapidly.

This in itself is a worthwhile outcome, but is enhanced by the denial to potential weapons proliferators of access to the plutonium remaining in spent reactor fuel.

By using plutonium in reprocessed mixed oxide fuels, any remaining after use is rendered unsuitable for nuclear weapons applications by the depletion of its readily fissile isotopes, in the same way that uranium fuel becomes depleted in its fissile isotope U-235 in a power reactor. The total quantum of available and weapons-usable plutonium is thus substantially reduced.

Nuclear Weapons Technologies

The complexities and implications inherent in any discussion of the basic form, structure and vulnerabilities of the nuclear fuel cycle in power generation are reflected in the nature of nuclear weapons, the nuclear technologies which have produced them, and the threats posed by uncontrolled diffusion of nuclear weapons ever further around the world. It is no exaggeration whatever to state that nuclear weapons are unique in the history of human civilization and warfare, being many orders of magnitude more dangerous than all of their predecessors, and carrying the capacity ultimately to destroy the planet.

And yet, the technologies which have produced these engines of Armageddon are indistinguishable from those which have, for over half a century, promised (but only partially delivered) unlimited energy for the benefit of humanity. The resulting conundrum, in which *that which makes us stronger can also kill us*, lies at the heart of nuclear homeland security policy, as much for the rest of the world as for the United States itself. Those countries which aspire to the possession of nuclear weapons, and those which have succeeded in obtaining them, believe that these devices bestow a cachet and prestige in the international community which is not available to them by any other means. India and Pakistan fall into this category, as does North Korea. For other states, including Israel, nuclear weapons represent the ultimate deterrent against the designs of any enemy, including those armed with nuclear weapons themselves. Simply put, the governments of these states calculate that a potential adversary which is convinced that its aggressive use of a nuclear weapon will simply result in its own obliteration is extremely unlikely to take the course of national suicide. Thus, it appears, at least prima facie, that nuclear devices are indeed not usable as weapons in war, since no state possessing them will ever, under any circumstances, take action to use them aggressively.[6] While this is, by some measures, a position unsupported by deterrent strategies linked to conventional military capacities, it remains as a sub-text for all acquisition and deployment of nuclear explosive devices.

But what are the nature and effects of these unique devices? Nuclear weapons which rely solely on nuclear fission – the chain reaction at the center of nuclear power generation – are limited in their power, and thus their effects, by the size of

---

6 Nuclear Threat Initiative, Global Security Newswire, *British nuclear deterrent 'useless,' retired officers say.* Available online: http://gsn.nti.org/gsn/nw_20090116_9271.php. Accessed January 17, 2009.

their fissile material assembly and the capacity of conventional explosives to induce in it a chain reaction. In practical terms, this amounts to a maximum possible yield (measured as the explosive force equal to a given tonnage of trinitrotoluene or TNT conventional explosive) of around 500 kilotons, or one half million tonnes. An explosion of this magnitude would amount to a detonation around 38 times stronger than that which, using a fission weapon, devastated Hiroshima in 1945.

And yet an explosion of this magnitude from a relatively simple fission bomb, in which a critical mass of uranium-235 or plutonium-239 results in the runaway chain reaction described earlier, is quite small when compared to those available from Hydrogen, or thermo-nuclear, weapons.[7] The yield available from the H Bomb's basic Teller-Ulam design is essentially unlimited, the Soviet Union having in 1961 tested a hydrogen device with a yield estimated at 51 megatons, or million tons, of TNT equivalent. Combining the physical principles of both nuclear fission and fusion (the latter using the lightest, instead of the heaviest of materials along the Periodic Table of chemical elements), thermo-nuclear weapons represent the ultimate terror weapon. Even so, the effects of fission and fusion explosive devices are similar, and need to be understood in order to have any chance of comprehending the nature and scale of the risks involved in their use. Should these weapons come under the control of non-state groups or individuals, their deployment and detonation would permanently alter the nature and shape of international security, and thus of homeland security.

Since legitimate employment of fissile uranium and plutonium is almost completely limited to their peaceful use in power generation, diversion from that domain into nuclear weapons programs is the most serious and dangerous pathway for those states, such as North Korea, India and Pakistan, which aspire to be nuclear weapon powers. It can be noted at this point, however, that this is not the only way to accomplish nuclear weapons ownership, since most states possess minimally usable amounts of recoverable uranium, and may be able indigenously to develop or import the technical skills and materials necessary to fabricate weapons without a civilian nuclear power program.

The Nature of a Nuclear Explosion[8]

Any nuclear explosion will consist of some mix of the following immediate effects, depending on the size and type of weapon involved and the site of its detonation.

7 In the case of uranium, the spherically shaped critical mass needed to produce a chain reaction resembles a basketball; in the case of plutonium a large grapefruit.

8 Although many aspects of nuclear weapon design are highly classified, the effects of nuclear detonations in the atmosphere are well known and understood. This overview is based on: 1) Johnston, W.R., *Nuclear weapons effects – an overview*, available online: Accessed January 17, 2009; and 2) The nuclear weapons archive, http://www.johnstonearchive.net/nucleareffectsum.html *5.0 Effects of nuclear explosions*. Available

In general terms, all nuclear explosions transfer their energy to the surrounding environment in three forms: blast, thermal energy and nuclear energy.

Fireball

A fireball appears as gamma rays superheat the air and the remains of the nuclear device itself. It quickly rises, forming a mushroom-shaped cloud as it cools in the upper atmosphere. The iconic mushroom cloud consists of condensed particles of water and vaporized materials from the explosion, and possibly ground debris, and forms a plume of radioactive material when driven by prevailing winds. These particles will eventually reach the ground as radioactive "fallout".

Flash

The fireball, at initial temperatures of tens of millions of degrees centigrade, emits large amounts of ionizing radiation in the form of x-rays, gamma rays and neutrons, as well as intense visible, infrared and ultraviolet light and other thermal radiation, for a short time as it develops. These flash effects will cause instantaneous heating and spontaneous ignition of combustible material at relatively close ranges, as well as burns and eye injuries to people at ranges out to many kilometers from the explosion site. Large scale fire conditions, including firestorms, are possible in vulnerable areas.

Ground shock

Shock waves created by the severe compression of the surrounding atmosphere by the rapidly expanding fire ball are intense, the over-pressure being capable of destroying most solid structures at ground level, and even hardened underground structures.

Electro-magnetic pulse

Ionizing radiation from nuclear explosions at altitudes above 30 kilometers will produce an intense electro-magnetic pulse capable of severe disruption or destruction of sophisticated electronic and information technology-related equipment, both civilian and military, over a very wide area.

This brief overview of the nature and major effects of nuclear explosions is intended only as a reminder that, when decision makers engage in the field of nuclear armaments, they must do so in full knowledge of the existential dangers

---

online: http://nuclearweaponarchive.org/Nwfaq/Nfaq5.html. Accessed January 17, 2009; and 3) Federation of American Scientists, *Special weapons primer: nuclear weapon effects*, available online: http://www.fas.org/nuke/intro/nuke/effects.htm. Accessed January 19, 2007.

which errors and unintended consequences may generate. When entering the realm of nuclear weapons – their development, deployment, use as threat or deterrent – whether as an independent observer, nuclear policy maker, proliferator or agent of non-proliferation, all actors step into the unknown. Having crossed this threshold, it is in the nuclear domain that all aspects of national security policy development and application can be seen in their sharpest relief. In pursuit of securing the homeland, no policy agent or other actor can afford to lose sight of the cost of nuclear security failure.

Scheme of the Book

Having set out in this Introduction the scope of its discussion, the book's subsequent chapters develop in greater depth its assumptions, rationale, theoretical foundations, evidential approach, analytical strategy, findings and recommendations.

Chapter 2, "Securing the Homeland, Securing the World" aims to lay out the parameters of the homeland security paradigm within the context of a broad, partially speculative discussion about the evolving nature of national security on a global scale. In doing so, it uses two linked approaches. First, it employs the current homeland security strategic settings of the United States as an identifier and benchmarked context for the evolving nature of national security at the regional and global scales.

Furthermore, it builds on the previously noted structured sets of nuclear-relevant states (established, new and aspirant) together with the fourth, non-state, focus of analysis. In this way, the reader is introduced to the general shape and direction of evolving threats to homeland security, and more specifically to existential threats based on nuclear proliferation.

Second, by linking state and regionally-based national security regimes in complexly contingent ways across the field of enriched uranium/plutonium global security issues, the chapter highlights the need for an internationally integrated and coherent approach to the safeguarding of nuclear fuels. More specifically, it will undertake a review of the current state of play in regard to the global security of natural uranium, enriched uranium and also plutonium. Here, it will provide an initial "snapshot" of data and trends in relation to the current location, production, storage, transportation, safeguarding, export and use of these three nuclear fuel types around the world.

In addition, it will provide a general assessment of the nature and scale of contemporary and emerging deficiencies in nuclear fuel security. The risks attendant on implementing a set of alternative and complementary national, regional and global policy initiatives aimed at increasing nuclear fuel safety will be balanced in this assessment against the risks of failing adequately to tackle the problem.

The chapter's final section will prepare the way for Chapter 3's discussion of relevant international institutional frameworks by noting the difficult task of coordinating initiatives across individual countries, alliances, regions, and the entire international community.

Chapter 3: "The International Institutional Framework" emphasizes the need for strong and durable legal and political frameworks deployed in support of efforts to safeguard nuclear fuels. It uses a "levels of analysis" approach to underscore the complexity and redundancy of policy and action which is a feature of the international nuclear non-proliferation regime generally, and which is increasingly echoed in the sub-set of global nuclear fuel security measures.

The chapter uses international law on the use of force as its organizing analytical framework. The sources of international law (custom, treaty law, accepted legal principles and legal scholarship) provide a structure for discussion of the wide range of law-based non-proliferation measures now in place. Within that broad frame, the effectiveness of international law in fostering novel approaches to nuclear fuel security (such as internationally-based enrichment and re-processing) is tested.

The chapter then places the challenges of nuclear fuel security within the sweeping spectrum of the nuclear non-proliferation regime. It examines, for example, the role, successes and deficiencies of the Global Initiative to Combat Nuclear Terrorism, the Proliferation Security Initiative, and the G8 Global Partnership Against the Spread of Weapons and Materials of Mass Destruction in responding to the nuclear fuel security conundrum. The corollary of this exercise is a set of initial proposals for reassessment and reform in cases where the regime's nuclear fuel security performance has been found wanting.

Chapter 4: "Assessing Security Threats Inherent in Nuclear Fuels" provides a comprehensive recapitulation of the general nature of the suite of nuclear fuel security threats. From this base, which incorporates the empirical data presentation and background analysis thus far undertaken, the chapter looks closely at each threat element with the aim of assessing the relative urgency of treatment demanded by each.

This exercise is necessary in order to establish the patterns of resource allocation and time scales within which a comprehensive program of enhanced global nuclear fuel security measures must be undertaken. In addition, perceptions of the strengths of the range of possible threat scenarios is diverse, a fact which itself demands that the greatest care be taken by policy planners and decision makers, especially in the arena of international nuclear negotiations. The analysis developed in this chapter is offered as a broadly based analytical resource to assist such individuals and advisers in their policy planning and implementation strategies, whether in the legal/United Nations sphere, or at the political negotiation level.

It is also at this point that the chapter emphasizes past, current and future nuclear fuel-based security threats emanating from non-state illicit activity. In this context, the history of the once globally active and clandestine nuclear technology proliferation network of the Pakistani scientist A.Q. Khan provides a startling and egregious example.

Chapter 5: "The Historical Record: Failure and Success" explores the historical landscape of success and failure in the sixty-year-old field of securing the nuclear fuel cycle. In particular, it looks at the record of deficiencies apparent in safeguarding the nuclear fuel stages of the cycle, and the consequences that have followed. The intention here is to highlight a range of experiences of both successful and failed global and regional policy formulation, interpretation and application, as a strategy for avoiding the pitfalls of the past while reinforcing success.

Examples helping to illustrate this process include North Korea's initiation of a weapons development program during the mid-1950s. A small case study is provided of Pakistan's protracted move to acquire nuclear weapons technologies and materials. The Pakistan example is especially highlighted in view of that state's long-term animus towards its sub-continental neighbor, rising power and nuclear armed state, India. The uneven and often inadequate global response to nuclear weapons proliferation, especially among the five nuclear weapon states, is emphasized in terms of opportunities lost and the continuing proliferation consequences flowing from the diffusion of nuclear technologies.

Chapter 6: "Strategic Security and Enriched Uranium" engages in a wide ranging examination of the agency of enriched uranium in helping to shape the global strategic landscape. The most telling factor, to be examined in depth, is the "pull factor" of the nuclear fuel cycle's capacity to produce the enriched uranium (and the plutonium) which makes feasible a clandestine nuclear weapons development program. For some states whose governments believe that there is, on balance, more strategic advantage to be gained than lost from such a program, the temptation to divert and enrich uranium for use in a weapons program may become too great to ignore. Iran will be used to illustrate this general point, and will also be examined in terms of its putative motives in setting up large scale uranium enrichment operations. Central to these considerations are Iran's importance as a major regional power in the Middle East, as well as the growing nuclear threat it is now perceived to pose in Europe. Iran's expected move towards weapons manufacture, when added to its demonstrated ballistic missile capacities, has been a pivotal factor in the former Bush Administration's proposed institution of a US ballistic missile defense shield based in the Czech Republic and Poland. Israel's reaction to these developments further complicates the regional strategic calculus.

The chapter continues by discussing the response of an increasingly strong and confident Russia to these developments. What Moscow perceives to be a pattern of steady encroachment and confrontation from the North Atlantic Treaty Organization (NATO) in Eastern Europe lends a further layer of strategic complexity. Russia itself has willingly participated in a long process by which much of its discarded weapons-grade highly enriched uranium (HEU) has been converted through down-blending for use in American nuclear power plants. This element of the enhanced global security of enriched uranium is now coming to an end, but should be regarded as a relatively rare success story in a noticeably barren field.

Chapter 7: "Nuclear Terrorism and Enriched Uranium" argues for the position that the world's nightmare scenario of active nuclear terrorism is a definable and present danger. However, its analytical strategy is to clarify and carefully assess the suite of threats which attach to, and construct, the ultimate security threat. First, the chapter reviews the evidence supporting claims in respect of terrorist group access to nuclear technology and materials, including enriched uranium and plutonium. This discussion is divided between access through the covert action of states, or empowered individuals within them (such as A.Q. Khan in Pakistan) and access afforded through false end-user and other black market transfers, often undertaken with the knowledge and active assistance of legitimate Western corporations.

Second, the chapter reviews the nature and scope of the nuclear terrorist threat created through access to nuclear technologies, contrasting this with the acquisition by terrorists of complete and operational nuclear devices. Finally, it reviews the range of responses to these perceived threats, their efficacy, and their capacity to render nuclear terrorism difficult or impossible to accomplish.

Chapter 8: "Old and New Strategies for Non-proliferation Progress" comprises a comprehensive review of the set of continuing and emerging global non-proliferation strategies. It places the issue of independent international control over the enrichment of uranium and the availability of plutonium for use in nuclear explosive devices within this broader context.

Indicative issues to be addressed here include the Global Nuclear Energy Partnership and its alternatives, such as an IAEA-sponsored international fuel bank proposal; US/India nuclear cooperation; transformation in the size and security of the US nuclear stockpile; proliferation-resistant reprocessing of spent nuclear fuel; and the down-blending and use of surplus Russian highly enriched uranium for power generation within the United States.

The chapter develops an integrated approach towards these and other non-proliferation efforts in the belief that all possible and viable options must be canvassed, tested and applied in the cause of countering the spread of nuclear technologies, of which nuclear fuel security is an integral component. It also develops the general point that strategies which are successful in slowing the rate of nuclear diffusion will also have the effect of reducing the likelihood that terrorist organizations will obtain, and then use, explosive nuclear devices.

Chapter 9: "The Future: Alternative Policy Pathways for the Twenty-first Century" is a partially speculative examination of possible pathways of change in the nuclear non-proliferation mix – and more specifically in respect of new developments in the holding, manufacturing, transportation, security, import or export of enriched uranium or plutonium. As the erosion of the nuclear non-proliferation regime continues apace, the chapter advocates the construction by the United States and its allies of a network of durable regional and global coalitions of states with the capacities and political will, under the aegis of a revitalized International Atomic Energy Agency, to cooperate in effective ways to deny nuclear fuels to states determined to continue towards nuclear weapons ownership. These coalitions should be constructed not only to address notable nuclear "hot spots" such as North Asia, the Middle East and the Indian subcontinent, but should attempt to anticipate and resist new and illicit nuclear policy directions among states.

The final chapter will engage in a discussion about the need, which appears now to be underway with the guidance of the Obama Administration, for a change of emphasis across the policy field that can be summarized as a move from coercion to diplomacy. More specifically, there appears to be a greater chance of effective control over nuclear security issues, and policy responses, which are predicated on collective rather than unilateral action at the global level of analysis. Furthermore, progress is more durable when based on dialogue and integration, rather than isolation and confinement. The ruling principle should therefore be "trust but verify," combined with a far more rigorous approach to issues such as voluntary export control regimes focused on dual use technologies. Finally, observations will be made on the seemingly eternal conundrum that nuclear proliferation breeds insecurity, even as insecurity breeds enhanced nuclear proliferation.

Chapter 10 will, however, take a cautiously optimistic view of the chances for success of emerging nuclear non-proliferation strategies, and especially the move to take far closer control over the provision of nuclear fuel for power generation. The obvious implication here is the removal of any viable reason for states with bona fide and defensible aspirations for a nuclear power industry to indulge in the enrichment of uranium to weapon-usable levels, or to obtain plutonium suitable for the same purpose. Such a move also holds out the possibility of further promoting the use of nuclear energy as a low carbon emission alternative to base-load fossil fuel power generation by offering to states incentives for their irredeemable abandonment of nuclear armament ambitions.

A Comparative Analytical Approach

There is a large array of policy advocacy, decision making and implementation options available to agents active within the nuclear security field, and none more so than in respect of the securing of nuclear fuels from diversion towards illicit use. Of these options, some will have universal application, such as the recent proposal by the Nuclear Threat Initiative to institute a fuel bank approach to the placement of all nuclear fuels under the control of the International Atomic Energy Agency.[9]

Some, on the other hand, will be contingent on a range of independent variables, such as the relevant nuclear assets which individual states need to protect. In this category are assets such as new proliferation resistant science and technology

9 Nuclear Threat Initiative, NTI News, *Sam Nunn applauds EU contribution to IAEA fuel bank*. Available online: http://www.nti.org/index.php. Accessed January 22, 2009.

developments, and operational nuclear fuel enrichment and reprocessing facilities. Other factors to be taken into account in assessing the validity and strength of specific security measures include the calculus of a state's strategic and regional security settings (in, for example, the India – Pakistan and North Korea cases). Yet others will be contingent on the nature and durability of a state's alliance relationships, one example here being provided by Japan's increasingly active regional security presence in the context of its alliance relationship with the United States. Thus, it is essential to consider the contingent and variable nature of the processes by which national security threats are perceived, and policy solutions proposed.

It is clear, then, that the identification and application of nuclear fuel security measures among the many nuclear-interested states across the globe will differ to the extent that specific, and often unique political, diplomatic and other national and international criteria will be conceived, considered and applied within a state's nuclear fuel security policy mix. This is not to suggest that states will develop such policy without taking full account of their obligations under international law and the international system of Westphalian states. Rather, it is to underline the complexities which are inevitable in seeking to understand the global shape of the relationship between nuclear fuel security and homeland (essentially "national") security. Thus, the analytical context in which the book's structure, centered on the three sets of state-interested countries – the established nuclear states, newly nuclear countries, and aspirant or clandestine nuclear nations – and a fourth group of non-state agents such as terrorist groups and science and defense communities, divides the breadth and depth of the nuclear fuel security challenge.

Chapter 2 begins by using the parameters of the US homeland security paradigm, understood in a global security context, as a benchmark for an initial discussion about the evolving nature of national security at the regional and global scales, with emphasis on national nuclear security.

Chapter 2
Securing the Homeland, Securing the World

Homeland Security: A Vision

Chapter 1 briefly examined the paradigm of Homeland Security as expressed in the Preamble to the United States Constitution and the US Homeland Security Act of 2002. While the Preamble establishes the purpose of the American state and nation in terms of national defense, general welfare and durable liberty, the Homeland Security Act determines government functions in three distinct domains:

- *Homeland Security*: protecting the common good domestically against attack;
- *General Welfare*: advancing the common good domestically;
- *National Security*: protecting and advancing the common good internationally.

Of these, Homeland Security and National Security have the most direct relevance to this analysis. Homeland Security was further defined as "the deterrence, prevention, and preemption of, and defense against, aggression targeted at US territory, sovereignty, population, and infrastructure as well as the management of the consequences of such aggression and other domestic emergencies". The Homeland Security paradigm was in turn further defined within two sub-sets:

- *Homeland Defense*: the deterrence, prevention, and preemption of and defense against direct attacks aimed at US territory, population, and infrastructure;
- *Civil Support*: Department of Defense (DOD) support to civilian authorities for natural and man-made domestic emergencies, civil disturbances, and designated law enforcement efforts.[1]

In the *National Strategy for Homeland Security* document issued by the National Security Council in October 2007, former president George W. Bush addressed his Administration's vision of the Homeland Security paradigm as conceptualized, translated and applied in response to the need to strengthen America's preparedness for both natural and man-made disasters[2]. Referring to a suite of threats ranging

1 See Chapter 1, pp. 8–10.

2 George W. Bush, Homeland Security Council, *National strategy for homeland security, October 2007*, Washington, DC, The White House.

from disease pandemics to nuclear terrorism and the effects of natural disasters such as that of Hurricane Katrina on New Orleans, he called on Americans to contribute towards enhanced national preparedness. Emphasizing the need to cooperate with multinational threat responses, he thereby also acknowledged the global scope of the single most serious threat to US homeland security – the threat of nuclear terrorism built on failure to curb the global spread of nuclear weapons, technologies and materials.

In guiding, organizing and unifying America's homeland security effort, the national strategy document provides a common framework focused on four goals:

- prevent and disrupt terrorist attacks;
- protect the American people, critical infrastructure, and key resources;
- respond to, and recover from, incidents that occur; and
- continue to strengthen the foundation to ensure long-term success.

As a broad security taxonomy, the general scheme of US homeland security provides a benchmark from which to conceptualize the evolving nature of national security on a global scale. The national strategy for homeland security, however, offers the following, more focused, definition:

> Homeland security is a concerted national effort to prevent terrorist attacks within the United States, reduce America's vulnerability to terrorism, and minimize the damage and recover from attacks that do occur.[3]

The strategy goes on to discuss the structure of threats to the American homeland, framing them in terms of terrorism, catastrophic natural disasters, and catastrophic accidents and other hazards. It names the al-Qaeda terrorist network as the most dangerous and persistent manifestation of terrorism, operating in cooperation with regional terrorist groups and affiliates around the world.

Most significantly, though, it warns against any complacency over al-Qaeda's continuing efforts to acquire weapons of mass destruction, including the acquisition of chemical, biological, radiological and nuclear materials and technologies. It is also worth noting that the homeland security strategy also points to threats posed by other terrorist groups, such as Hizballah in Lebanon. It is thus sufficiently pragmatic to allow acknowledgement of the possibility of home-grown terrorism stemming from the radicalization of Muslim American groups and activists.

In summing up a vision and strategy for the enhanced security of the American homeland, the strategy notes the durable aims of terrorism against the United States: actions focused on prominent political, economic and infrastructure targets designed to produce mass casualties, visually dramatic destruction, significant economic damage, fear in the population, and loss of confidence in government.

---

3 Ibid.

In response, the strategy calls for the concerted and coordinated efforts of all arms of government – Federal, State, local and Tribal; the private and non-profit sectors; and regions, communities and individual citizens.

However, it also enlists the cooperation of America's allies and international partners in the fight against terrorism. In so doing, it calls into service all the instruments of America's national power and influence, including its diplomatic, informational, military, economic, financial, intelligence, and law enforcement capabilities. The most important of all these categories of pledge is, arguably, its commitment to cooperate at the international level to rein in, and ultimately destroy, the burgeoning terrorist threat. President Obama's new administration can be reasonably expected to take early steps to enhance America's international cooperative efforts beyond those put in place by his predecessor, and particularly so in response to identified and evolving nuclear threats.

There is no doubt that, in a process similar to the emerging US Homeland Security paradigm, many states have revisited what they perceive to be their national security goals, as the field of security is itself renovated to reflect an era of rapid threat development.[4] First among many threats will, for many states, be the real threat of nuclear terrorism which is based on both qualitative and quantitative assessments of the risks of active nuclear theft and terrorism, as well as the vulnerabilities of stockpiles of nuclear warheads, materials and technologies.

In regard to the specific focus of this analysis – primarily the security of nuclear power generation and warhead fissile materials in the context of general nuclear proliferation – the pace of change in the global re-ordering and re-prioritizing of national security goals provides an opportunity to test the validity of the nuclear state groupings noted in the Introduction. Simply put, does the structuring of nuclear-interested states into sets of *established, developing and nuclear capable*, and *aspirant or clandestine* categories help in understanding what needs to be done to safeguard nuclear fuels such as enriched uranium against illicit diversion? Furthermore, can the range of non-state nuclear agents (including not only terrorists and their organizations, but also the nuclear science community, non-governmental activist groups and nuclear power corporations) contribute further in building a comprehensive picture of the complexities of nuclear fuel security?

This seems, prima facie, to be a testable proposition, and will be undertake in this chapter through an overview of the suite of current and future nuclear fuel security issues at various level of analysis, including the sub-state, state, regional and global levels, and across the range of states and other actors engaged in

4 The Copenhagen School of Critical Security studies is an attempt to theorize the nature of "security" in both broader and deeper contexts, often with resonance for ideals of Human Security. It uses a critical approach to answer the question "Who can *do* security on behalf of whom, or what?" Its leading exponent has been Barry Buzan who, with colleagues, has challenged traditional security paradigms in seeking to understand the nature of security at the beginning of the twenty-first century. Buzan, B., Waever, O., and de Wilde, J. (1997) *Security: A new framework for analysis*. Boulder, CO: Lynne Rienner.

significant nuclear activities. In this way, linkages and interdependencies across the various operational (as well as diplomatic and pragmatic) levels will become clear, as will the enormity and complexity of the non-proliferation tasks involved.

The Scope of Nuclear Fuel Security Issues

The first task in understanding the importance and inherent difficulties to be overcome in confronting issues of nuclear fuel security is to assess the scale and diversity of specific threat scenarios, and the contexts within which they must be faced. The United States will continue to drive much of the global effort currently underway to rein in threats which have within them the capacity to produce the ultimate nightmare – terrorist acts involving nuclear weapons in the United States homeland, or the territories of other states. Nevertheless, the plain facts are that in much of the nuclear world (including the United States itself) nuclear security continues to be found wanting. Large stockpiles of nuclear weapons, with their highly enriched uranium (HEU) or plutonium fissile cores in place, continue to be poorly guarded in Russia, although progress in this direction is now occurring. Across the globe, many research reactors and critical nuclear research assemblies, often with rudimentary or insufficient security in place, use HEU or plutonium. Many of these are on Russian territory, some are within the United States homeland. All are potentially vulnerable to determined attack by individuals terrorist and their organizations.

Again, Pakistan, as a newly nuclear state, faces serious threats to its heavily guarded nuclear devices and materials, not the least of them stemming from the proximity of a resurgent al-Qaeda, based as it is in Pakistan's barely governable northern tribal areas along its border with Afghanistan.

Yet another source of nuclear fuel insecurity stems from the growing use of plutonium in civilian nuclear power generation, creating larger than necessary quantities of fissile material which is potentially available for explosives applications. Many other nuclear fuel security risks will emerge as this analysis proceeds.

It must be acknowledged here that it is not plausible to suggest that any known terrorist organization can currently aspire to unilaterally reach a capacity to create nuclear fuels and materials using only the resources available to it. A terrorist organization, for example, could not plausibly build, hide from view and operate a full scale centrifuge cascade of the size and cost necessary to enrich natural uranium, even if it could find a sufficiently large source of uranium hexafluoride gas with which to charge it. Nor could it initiate or complete the many technically difficult processing steps necessary to create finished HEU ready for bomb use. Again, the generation of bomb-ready plutonium-239 (the most readily usable isotope of plutonium) requires equally or more difficult technical and other capacities which must surely be beyond the reach of terrorist groups unassisted by the resources of a sympathetic and nuclear capable state.

This places the greatest burden of nuclear fuel security squarely on keeping HEU and plutonium out of the hands of terrorists through theft or other forms of illicit diversion by third parties. Individuals or groups of "insiders" within state nuclear establishments may well be motivated by greed, ideology or the temptations of political power to allow clandestine access to their country's nuclear materials, technologies or, indeed, complete nuclear weapons to aspiring nuclear terrorists.

There are a number of strategies available to successfully counter this and other threats based on access to fissile materials, and we will return to them below in greater detail. However, some of the most obvious and effective include efforts to:

- consolidate states' diverse weapons and other fissile materials holding sites into fewer, more easily defended sites;
- interdict nuclear smuggling operations;
- monitor and deter states from deliberate nuclear materials transfers to terrorists;
- obstruct and impede the clandestine recruitment of nuclear experts;
- further reduce global nuclear stockpiles;
- end new production of weapons-grade fissile materials;
- and institute global monitoring of all nuclear stockpiles.[5]

If each of these goals were attained, at least to the extent of reducing attendant risks to manageable levels, the overall threat of successful nuclear terrorism would be severely curtailed. Yet the dilemma of the defender comes forever to the fore: an attacker is often able to repeat his attack until he achieves success, while a defender will only ever fail once. Should that occur in the nuclear context, the world would take a step into the unknown in which the Westphalian international system of states would be severely challenged, and may be irreparably damaged. In order to ensure that nuclear fissile materials will never be used by any actor in any circumstances, an essential first step is to survey the current position in regard to the disposition of fissile materials around the globe, and the nature and strength of security threats surrounding them.

It is also appropriate to reiterate at this point that the thrust of this discussion focuses primarily on highly enriched uranium (which for weapons use is usually enriched in its fissile isotope U-235 to around 90–95 per cent) and plutonium-239, each in a form capable of being stolen or otherwise diverted from legitimate uses. As noted earlier, it is *not* primarily concerned with low-enriched uranium (LEU) at 3.5–5 per cent U-235 enrichment levels and suitable for use as fuel in many large power reactors. Equally, the 1 per cent plutonium component present in the spent fuel rods of many power reactors fueled by HEU is not available for further use

5 Matthew Bunn, *Securing the bomb 2008* (Cambridge, Mass., and Washington, DC: Project on Managing the Atom, Harvard University and Nuclear Threat Initiative, November 2008). At Executive Summary, pp. v–x.

until it has been separated from the spent uranium and fission products through nuclear fuel reprocessing. Even then, the separated plutonium would require enrichment in its plutonium-239 isotope to over 90 per cent in order to render it useful in a crude fission explosive device, a difficult and dangerous task.

For these reasons, terrorist organizations can no more plausibly claim the capacity to exploit the vulnerabilities of civil nuclear power programs – at least in terms of fuel sources – than they can in manufacturing uranium or plutonium in forms usable in improvised nuclear devices (INDs). For the nuclear terrorists, theft or other forms of misappropriation of metallic HEU or separated plutonium are the only plausible pathways to their mounting serious nuclear weapons threats.

Having noted the scope of nuclear fuel security issues, the field of nuclear-interested states can be laid out.

Categories of Nuclear-interested States

Established nuclear states

This category of established nuclear states includes the five nuclear weapon states recognized within the terms of the 1970 Nuclear Non-proliferation Treaty (NPT). Headed by the United States, it also includes the Russian Federation (as successor state to the Soviet Union), the United Kingdom, France and China. Although the United States and Russia surpass the other three nuclear weapon states in nuclear weapons holdings by very wide margins, all five possess credible and operational nuclear weapon strike forces. Furthermore, all five have extensive (and in China's case, actively growing) commercial nuclear power industries, and most retain active research reactors or critical research assemblies using highly enriched uranium (HEU).[6] In addition, both the United Kingdom and France hold substantial stocks of separated civilian plutonium.

Also in this category are a number of states with mature commercial nuclear power industries and HEU-fueled research reactors, but with no holdings of nuclear weapons. With some 280 research reactors operating in 56 countries, many are found in states with long nuclear histories, and the likelihood of reasonably effective levels of security against nuclear materials theft. They include Japan and NATO members of the Western Alliance such as Canada, Belgium, Germany and the Netherlands.

---

6 Research reactors are a lot smaller than commercial power reactors, and are used primarily as "neutron factories" for research and training, materials testing, and the production of radioisotopes for medicine and industry. They comprise the smallest category of nuclear reactors, are found typically on university campuses, and many are currently being converted to use low-enriched uranium (LEU) as fuel. World Nuclear Association, Information Papers: *Research Reactors,* London, UK. Available online, http://www.world-nuclear.org/info/inf61.html. Accessed January 31, 2009.

Developing states with nuclear weapons

The three states in this category are India, Pakistan and North Korea. India, which has not signed the Nuclear Nonproliferation Treaty (NPT), has of necessity developed a strongly indigenous nuclear power sector. As a result of its determination not to sign the NPT, as well as its 1974 detonation of a self-described "peaceful" explosive nuclear device, and several more nuclear weapons tests in 1998, India was for thirty-four years excluded from international trade in nuclear plant or materials. However, in 2008, India concluded nuclear cooperation agreements with the United States which included subsidiary compacts with France, the International Atomic Energy Agency (IAEA) and the Nuclear Suppliers Group of nuclear materials exporting countries (NSG).

Having finally been brought into the nuclear fold, there are now greater prospects for cooperation from the world's largest democracy on enhancing the physical security protection systems for its nuclear facilities. Although India's protection systems are believed to have been adequate in the past, its resources are still considered insufficient to robustly meet the need for better protection against nascent and increasingly aggressive terrorist threats.[7] Given India's possession of nuclear weapons with HEU cores, as well as separated plutonium for use in large scale power reactors, and uranium enrichment and reprocessing facilities, its current refusal to cooperate with the United States National Nuclear Security Agency (NNSA) on nuclear security issues reflects internal political difficulties for the Indian Government, rather than technical or other barriers.

Pakistan presents a separate and distinct set of nuclear fuel security challenges. Far less politically or economically stable than India, it continues to host the al-Qaeda terrorist organization within its northern borders. In a nation riven by official corruption and *jihadi* Islamic extremism, nuclear security threats emanate from sources both within and external to its nuclear establishment. It is also the home of the notorious A.Q. Khan global black market nuclear technology network, the discovery of which has, ironically enough, provided the impetus behind considerable Pakistani efforts to strengthen its nuclear security measures.[8]

Nevertheless, US efforts to help reinforce Pakistan's nuclear security and accounting methods is hampered by a significant and widespread degree of suspicion over America's motives, including a belief that Washington is attempting to seize Pakistan's nuclear arsenal – thus rendering it defenseless against its Indian adversary. In all of this, there remains the small but real possibility of the collapse of the Pakistani state, *coup d'etat* by jihadists with Taliban links, or concentrated extremist attacks on nuclear facilities – with or without collusion by sympathetic senior military officers.

---

7 The US National Nuclear Security Agency has responsibility within the US Department of Energy (DOE) for working with both India and Pakistan on measures to improve the security of their respective nuclear establishments. See Footnote 5, at p. 23.

8 Ibid.

The third member of this group, North Korea (or the DPRK) does not present a significant threat in terms of nuclear fuel security, *per se*. Its Yongbyon nuclear facility has produced a small stockpile of around 30 kilograms of separated plutonium, while its stock of nuclear weapons can only be extremely small. The Six Party nuclear negotiations process undertaken between the DPRK, South Korea, the US, China, Japan and Russia, which were originally convened in reaction to North Korea's 2003 withdrawal from the NPT, have met with intermittent and limited success since their institution, although in 2007, Pyongyang agreed to shut down its nuclear facilities in exchange for oil fuel aid and diplomatic recognition by the United States and Japan. In any event, the only plausible threat from North Korea's fissile materials stocks comes from the possibility of the collapse of the Kim dictatorship, leaving a military and political vacuum ripe for exploitation by individuals or groups able to take advantage of social, political and economic collapse. At the time of writing, the Six Party talks are again in limbo as a result of North Korea's angry reaction to UN criticism of its April 2009 launch of a long-range missile over Japan and towards the North Pacific Ocean, under the guise of a satellite launch exercise.

Finally, a small sub-set exists in this category: developing states without nuclear weapons, but with small nuclear infrastructures and substantial holdings of HEU, mostly as fuel for use in research reactors. Often with inadequate physical protection – especially when located on university campuses – the largest stocks in this category are held by South Africa (at its large scale Pelindaba nuclear facility), Belarus and Ukraine.

Aspirant and clandestine nuclear states

Iran and Israel qualify as aspirant and clandestine states, respectively. Iran's nuclear ambitions have been of concern in the West for some years, growing increasingly serious from the late 1990s. Iran is driving forward in its quest to establish all aspects of the nuclear fuel cycle, ostensibly with the aim of developing an indigenous civilian nuclear power industry. Its plans include uranium conversion and enrichment facilities, and heavy water production and fuel fabrication plants. Notwithstanding the protestations of the Iranian government that this activity is merely in preparation for the construction of a nuclear power industry, if it succeeds in these plans Iran will soon present a clear nuclear weapons threat to the Western Alliance and the Middle East as it seeks political, religious and military dominance in its region. In the context for these discussions, however, the primary nuclear threat is constituted by the widespread perception within Iran's region and beyond that it may be willing, having obtained the capacity to build nuclear weapons, to export nuclear fissile materials, technologies and skills to terrorist groups and others seeking to use them against the United States and its allies.

Israel, a close and long-standing ally of the United States, has always denied the existence of what is widely believed to be a significant holding of nuclear weapons (of both plutonium-fueled fission and thermo-nuclear types) of between 100 and

200 weapons. The 1986 revelations by Mordechai Vanunu, a former employee at the Negev Nuclear Research Center reactor near Dimona, of its existence and plutonium production capacities, have never been successfully countered by Israeli governments. Whatever the truth, there appears to be no realistic chance of terrorist linked diversion of plutonium or HEU, given the extraordinary military security measures in place to protect the Dimona facility.[9] However, the real intent of Iran as an aspirant nuclear state, and the existence or otherwise of a clandestine Israeli nuclear weapons program may well be tested as Israel grows ever more concerned about Iran's nuclear capacities and intentions toward her, and her Gulf neighbors.

Non-state nuclear actors

Most legitimate non-state actors in the nuclear field do not present credible nuclear fuel threats to states and their populations through consciously undertaken actions, not least because they are usually under considerable scrutiny by agents of the state or states in which they operate. Some are also subject to oversight by international governmental organizations and agencies such as the International Atomic Energy Agency and the Nuclear Suppliers Group. In this category are the nuclear power generation corporations, the many commercial corporations supplying nuclear related materials and technologies world wide, nuclear industry peak representative bodies, and the panoply of not-for-profit non-governmental organizations active in the promotion of, or resistance against, all aspects of the use of nuclear energy.

Although there is a long history of attempts by commercial suppliers of nuclear-related products and services to circumvent established rules on their international movement, as well as more recent attempts by states (for example, the US-led Proliferation Security Initiative) to counter such activities, it is probably impossible to completely eliminate the problem.[10]

A Fissile Material Global "Snapshot"

While a full account of the location, inventory, production, storage, transportation, safeguarding and use of highly enriched uranium and plutonium around the world is beyond the capacity of this analysis, it is possible to construct a comprehensive summary of current world practices in dealing with, and securing, potential nuclear

9 For a detailed assessment of Israel's nuclear capacities, see: Nuclear Threat Initiative, *Israel profile: Nuclear overview*, Available online at: http://www.nti.org/e_research/profiles/Israel/Nuclear/. Accessed February 3, 2009.

10 Strategic Documents, *Nuclear black markets: Pakistan, A.Q. Khan and the rise of proliferation networks*, The International Institute for Strategic Studies, London, Available online at: http://www.iiss.org/publications/strategic-dossiers/. Accessed February 3, 2009.

Table 2.1 Global plutonium and highly enriched uranium (HEU), assigned to civil or military stocks, end of 2004, in metric tons

Category	Plutonium	HEU	Total
Civil Stocks	**1,675**	**175**	**1850**
Power and Research Reactor Programs	1,570	50	
Russian and USMilitary Excess	102.5	125 (US only)	
Military Stocks	**155**	**1,725**	**1880**
Primary	155	1,250	
Naval and Other	–	175	
Russian HEU Declared Excess	–	300	
Total	**1830**	**1900**	**3730**[a]

Note: [a] Institute for Science and International Security, Global stocks of nuclear explosive materials: Revised September 7, 2005. Available online at: http://www.isis-online.org/global_stocks/end2003/summary_global_stocks.pdf.

weapon grade fissile materials. This is primarily a quantitative exercise, and is based on the most contemporary and authoritative sources currently available in the public domain.[11] It will be followed by a qualitative assessment of the deficiencies in nuclear fuel security which the fissile material "snapshot" reveals.

Although the five nuclear weapon states (the United States, Russia, the United Kingdom, France and China) hold the lion's share of weapons-grade fissile material, over fifty countries now possess more than five kilograms each of weapons-usable material. India, Pakistan, Israel and North Korea, all nuclear weapons states, also hold significant stockpiles, while developed but non-nuclear weapon states such as Belgium, Germany, Japan and Switzerland possess sufficient material for multiple nuclear devices. More than 40 countries also hold some 20 metric tons (mt) of HEU in approximately 130 operational nuclear research facilities around the world.

Broad estimates put the global levels of highly enriched uranium and plutonium stockpiles at around 1,830 mt (metric tonnes) and 1,900 mt, respectively.[12] Most of the plutonium is, as discussed earlier, unavailable for use in improvised nuclear devices because it is embedded with the highly radioactive fission products found in the spent fuel assemblies of civilian production reactors. It would be extremely

11 Sources used include, but are not limited to: the Report *Securing the bomb 2008* (See Footnote 5); the Institute for Science and International Security, available online at: http://isis-online.org/, accessed February 4, 2009; and United States Government Departments and Agencies such as the National Nuclear Security Agency and the Department of Energy.

12 Institute for Science and International Security, Global stocks of nuclear explosive materials: Revised September 7, 2005. Available online at: http://www.isis-online.org/global_stocks/end2003/summary_global_stocks.pdf. Accessed February 4, 2009.

difficult, costly and dangerous for terrorists to attempt to gain access to it. Civilian and military stocks of separated plutonium account for some 230 mt and 250 mt respectively, sufficient together for many thousands of nuclear weapons.

It is necessary to note here that these broad estimates are subject to considerable statistical error, due to the failure of Russia to declare the level of its civilian separated plutonium stocks. Nevertheless, the figures noted in Table 2.1 are useful as indicative data.

The Second Report of the International Panel on Fissile Materials (IPFM), an independent group of arms control and non-proliferation experts from sixteen countries, describes its mission as ".... to analyze the technical basis for practical and achievable policy initiatives to secure, consolidate, and reduce stockpiles of highly enriched uranium and plutonium." As a consolidation and interpretation of recent, relevant and credible statistical data, the quantitative element of this chapter's analysis will note several explanations and clarifications of the raw data on global weapons-grade fissile material, which is summarized in the IPFM's most recently published Report.[13]

More specifically, the Report notes the following data in respect of highly enriched uranium; separated plutonium; the consolidation of fissile materials in the US nuclear complex; progress towards nuclear disarmament and international monitoring in the nuclear weapon states; the future of nuclear power industries; Russia's role in the international nuclear fuel cycle; and environmental monitoring used to detect clandestine fissile material production.

Highly enriched uranium

As of 2007, the global stock of highly enriched uranium is estimated at somewhere between 1,400 and 2,000 metric tons. Russia has now down-blended over 300 mt of HEU to low enrichment levels, and shipped it to the United States for use in standard power reactors where it accounts for around half of US power reactor fuel needs. When this program ends in 2013, around 500 mt of surplus Russian HEU, enough for some 20,000 nuclear weapons, will have been blended down and used in the US nuclear power industry.

At the same time, the US has blended down or otherwise disposed of some 87 mt of its surplus HEU (although not of weapons grade) and plans to add a further 147 mt to this total over coming decades. It is a sobering reality, nevertheless, that after the completion of these HEU disposal programs, the United States and Russia will retain a combined total of between 600 mt and 1,200 mt of HEU for weapons use, enough for between 25,000 and 50,000 weapons. In addition, the US, the United Kingdom and Russia have substantial stocks of HEU for use as fuel in

13 International Panel on Fissile Materials (IPFM), *Global Fissile Material Report 2007*. Available online at: http://www.fissilematerials.org. Accessed February 8, 2009. At p. 1.

naval vessels (nuclear powered submarines and, in Russia's case, icebreakers). Used for weapons, this stock of HEU would fuel another 5,000 weapons.

Finally, the on-going efforts by the United States in its Global Threat Reduction Program (conducted by the US National Nuclear Security Agency) have been successful in removing large quantities of Soviet and American-supplied weapons grade HEU from research reactors around the world. More work is need to complete the program, but to date some fifty research reactors – fifteen of them in developing states with small nuclear programs – have been converted to use low enriched uranium (LEU) which cannot be used in nuclear weapons. HEU has been completely eliminated from the civilian nuclear programs of sixteen countries, including developed states such as Greece, Italy, Portugal, South Korea, Spain and Sweden. Russia, however, has no program to eliminate its HUE from the seventy research reactors it currently operates.[14]

Separated plutonium

Again using 2007 estimates, global stocks of *separated* plutonium – that is, plutonium in a form which is capable of weapons applications – stands at around 500 mt, split almost equally between civil and military applications. Less than one quarter of the 75 mt of plutonium-239 which remains in the annual output of spent power reactor fuel around the world is currently being reprocessed to separate the plutonium. Nevertheless, almost all current global plutonium stocks, both civilian and military, are usable in nuclear weapons. The bulk of these stocks exist, unsurprisingly, in the five declared nuclear weapon states, although Japan (with almost 40 mt), Germany and India also hold significant quantities within their civilian programs.

Russia holds from 120 to 170 mt of military plutonium for potential use in nuclear weapons, and has declared up to 50 mt surplus to its requirements, while the United States has declared some 45 mt as in excess of its nuclear weapons needs. Neither state, nor France, the United Kingdom and China, are likely to produce any new stocks of plutonium for military use from 2010. However, progress in the disposal of these surplus plutonium stocks has been slow. [15]

The picture is less positive for Israel, India and Pakistan, however. Each is almost certainly continuing to produce plutonium for military use which, in the cases of India and Pakistan, will receive boosts from the completion of new reactors free of safeguards under the aegis of the International Atomic Energy Agency (IAEA). India plans the completion of a high-power plutonium fast breeder reactor with a capacity to produce around 140 kg of weapons-grade plutonium annually, while continuing to build its capacity to separate plutonium from spent civilian reactor fuel. Pakistan is also building its weapons-grade plutonium production capacities, aiming to triple output from its unsafeguarded Khushab reactor

14 Ibid. At p. 3. Supra. At pp. 51–2.

15 Ibid. At pp. 13–14.

Table 2.2 Total operational nuclear weapons warhead arsenals, including stocks using highly enriched uranium and plutonium, in fission, boosted fission and thermo-nuclear (Hydrogen or fusion) forms, were estimated at the following values in 2007

Country	Nuclear warheads
United States	Approximately 5,000 with a 5, 000 reserve
Russia	Approximately 10,000
France	348
UK	185
China	Approximately 130
Israel	Approximately 100
Pakistan	Approximately 60
India	Approximately 50
North Korea	Fewer than 10[a]

Note: [a]International Panel on Fissile Materials (IPFM), *Global Fissile Material Report 2007*. Available online at: http://www.fissilematerials.org at p.8.

facilities. Meanwhile, Pakistan's reprocessing of spent civilian fuel continues apace, reinforcing widely held perceptions that it is moving towards a plutonium-based nuclear weapons stockpile in the face of very limited indigenous supplies of natural uranium.

It is significant that the Strategic Offensive Reduction Treaty (SORT) negotiated between the United States and Russia in June 2003 calls for a maximum of 2,200 deployed strategic nuclear warhead to be held by each party by December 31, 2012. However, SORT does not require elimination of non-deployed weapons, which would be available from 2013 for renewed deployment.

Categories of nuclear threat[16]

This "snapshot" of the state of nuclear security, emphasizing the security of the nuclear materials contained within them – both HEU and plutonium – will be followed by a review of the range of plausible strategies for securing global nuclear stockpiles.

16 Supra. At Note 5. The structure of nuclear threats discussed here is based on Chapters 1 and 2 of *Securing the bomb 2008:* "Progress in Reducing the Risk of Nuclear Theft and Terrorism: A Qualitative Assessment"; and "Securing Nuclear Stockpiles: Quantitative Indicators".

The ultimate aim of all responses to each category of nuclear threat is to ensure, to the extent feasible, that terrorist and their organizations or affiliates do not reach a point where they can realistically aspire to mount a successful attack on any target using nuclear means. More insidiously, the corollary of such a situation is a widespread *perception*, held by national political leaders, their advisers and citizens, that such a threat does in fact exists. A perception of this kind, built on more or less complete or reliable intelligence products, will inevitably color the nature of any threat response, especially when the lives of millions of citizens and the continued existence of of the state lie in the balance.

The broad sweep of nuclear threat categories may be expressed thus:

- Inadequate security and monitoring of stocks of weapons-grade fissile material (nuclear weapons, separated plutonium and HEU) worldwide
- Continuing production of HEU and plutonium by states around the world
- Recruitment by terrorists of personnel with nuclear-applicable skills
- Acquisition by terrorists of operational nuclear weapons
- Covert international trade in nuclear materials and technologies
- Continuing development of state-based nuclear weapons programs.

Each of these categories contributes significantly to the nature and seriousness of threats posed by nuclear terrorism, and we will examine each in turn.

Inadequate Security and Monitoring of Weapons-grade Fissile Material Worldwide

As noted earlier, there remain large stocks of weapons-grade, or weapons-usable, fissile materials in complete nuclear weapons, both plutonium and HEU, stockpiled for current or future military applications, and in some 130 civilian nuclear research reactors currently operational around the world. A significant amount of HEU is also used in a few states (dominated by Canada) in the production of isotopes for medical applications, while a fourth application is for maritime nuclear propulsion in the submarine and surface fleets of the nuclear weapon states.

As noted earlier, unseparated civilian plutonium also exists in relatively large quantities in the spent fuel of most power production reactors, where it is functionally immune from terrorist activity because of its incorporation with the dangerously radioactive fission products also present in spent nuclear reactor fuel. However, some states have for many years reprocessed both their own and foreign spent reactor fuel to separate out its plutonium for re-use, when blended with uranium, as mixed oxide (MOX) fuel in suitable reactors. Although the international market for nuclear fuel reprocessing has collapsed over the past few years (prompting the United Kingdom, for example, to abandon its domestic and international reprocessing operations) France and Japan have retained long-term plans to separate plutonium to feed their nuclear power industries using MOX fuel.

Japan has recently commissioned its very large reprocessing facility at Rokkasho to recycle plutonium for use in mixed oxide fuels at a cost of around US$20 billion. This plant is now accumulating separated plutonium at the rate of 20 kilograms per day. For technical reasons, no plutonium has yet been used in MOX fuels as planned, thus adding to Japan's stockpile of some 40 metric tons of plutonium and representing one of the most significant sources of separated plutonium in the world. This must also present a tempting target for terrorist attack.

Other developed states, such as the United Kingdom and Russia, continue to hold large stockpiles of separated plutonium in storage for possible future use. In all, some 20 metric tons of separated plutonium is generated globally each year through the reprocessing of spent nuclear fuels, a quantity sufficient for many thousands of nuclear weapons – including the crude gun type devices which some terrorist organizations may be working towards.

The International Atomic Energy Agency (IAEA) based at Vienna, Austria monitors all stockpiles of plutonium and HEU in states which are not declared Nuclear Weapons States under the terms of the Nuclear Non-proliferation Treaty. Even so, IAEA monitoring to account for the quantities of fissile materials held does nothing, in itself, to help secure those stocks from diversion. Furthermore, most stocks of weapons grade fissile materials held by the five NPT-declared nuclear weapons states (the US, Russia, the UK, France and China) are not subject to IAEA monitoring, there being no requirement in the NPT to mandate this. As a result, data on the nuclear weapon states' holdings of weapons-grade fissile materials relies on the declarations made by these five states, and their willingness to allow IAEA oversight and verification of the integrity of their stockpiles. Transparency in this respect can therefore only be effected by ad hoc, bilateral and other international agreements on information sharing. To this extent, echoes of Cold War suspicion and animus can still be heard in the reluctance of the nuclear weapon states to open all of their military nuclear facilities to the scrutiny of independent stockpile integrity and security screening.

As a general comment, *all* stockpiles of HEU and separated plutonium, in both civilian and military applications and security arrangements are, of course, more or less vulnerable to threats of theft or other kinds of diversion towards nuclear terrorism activities. Successes in enhancing the security of weapons-grade fissile materials, and further security challenges, will be noted below and discussed further in later chapters.

Continuing Production of HEU and Plutonium

In addition to the on-going production of HEU and plutonium by developed and established nuclear states, production is growing in relatively new and less politically or economically stable states. The non-governmental International Panel on Fissile Materials has reported that none of the five nuclear weapons states declared under the terms of the Nuclear Non-proliferation Treaty are believed currently to be producing highly enriched uranium. In fact, only Pakistan and India are known to be doing so, the former for use in nuclear weapons, the latter for marine nuclear propulsion and possibly for future weapons applications. Each is currently constructing additional advanced technology reactors which will enhance the levels of plutonium available to them for use in nuclear weapons over the next decade. Neither will be operated under the safeguards and monitoring regime of the IAEA, while India is also moving forward on the construction of two fuel reprocessing facilities for the separation of additional weapons-grade plutonium.[17]

Finally, in addition to the continuing annual production of some 20 mt of civilian plutonium noted above, Russia produces around 1.2 mt per annum of separated weapons-grade plutonium from its three operational plutonium energy production reactors, but is currently considering their shutdown.

Recruitment by Terrorists of Personnel with Nuclear-applicable Skills

This threat category is at once the most troubling in its implications and perhaps the most convincing of all in its likelihood. There are well recognized reasons to fear that it is only a matter of time before a nuclear device is used as an instrument of terror based on the ease with which a terrorist group, *having first obtained the necessary fissile material*, could fabricate a nuclear explosive device. A recent discussion of the ways in which a determined terrorist cell could accomplish this feat lends a distinctly practical quality to what appears, at first glance, to be firmly in the realms of fiction.[18] The plain fact is, however, that since the necessary materials, designs and manufacturing techniques are either readily obtainable or in the public domain, there exists no functional impediment to a terrorist group fulfilling its nuclear ambitions once they have obtained sufficient and suitable HEU or plutonium.

How so? Wherever in the world it is attempted, the minimum steps necessary towards building a complete and functioning nuclear device are essentially the same. Using the crudest of weapons technologies (as used in the gun-type uranium fission device known as "Little Boy" dropped on Hiroshima, Japan in August, 1945) a team of fewer than twenty individuals with standard appropriate skills could construct an operational explosive device within six months. Using a surplus light artillery barrel, fissile critical mass would be reached by firing a metallic HEU bullet into a hollowed target of the same material bolted onto the end of the barrel. The result would be an uncontrolled chain reaction and nuclear explosion.

17 Supra. At Note 14, pp. 11–15.

18 Peter D. Zimmerman and Jeffrey G. Lewis, *The bomb in the backyard*, November/December 2006. Available online at: http://flotilla31.com/News/Bomb_in_the_Backyard.pdf. Accessed February 14, 2009. For a more extensive analysis, see: Mueller, J., *The atomic terrorist: Assessing the likelihood*, Department of Political Science, Ohio State University, January 1, 2008. Available online at: polisci.osu.edu/faculty/jmueller/APSACHGO.PDF. Accessed February 14, 2009.

The skills required for success amount to technical proficiency in advanced vacuum casting of uranium, artillery rifle refurbishment, supercritical uranium mass assembly, reliable electrical engineering, and the design, construction and final assembly of the finished bomb. Standard machine shop equipment sufficient for uses common to many simple engineering tasks, none of which requires special licenses to obtain, is all that would be necessary to complete a workable device. Furthermore, surplus recoilless artillery barrels are readily obtainable, although legitimately so only with the correct license in the United States and Canada.

The completed device would have cost somewhere in the region of US$1.5 – 2 million without accounting for the cost of buying its fissile uranium core, and would be easily transportable by light truck. Once again, though, by far the most difficult element in this process would be obtaining the HEU or plutonium, the Achilles heel of the nuclear terrorist and the best hope for more coherent and coordinated security strategies to deny the terrorists what they crave.

Acquisition by Terrorists of Operational Nuclear Weapons

Could a terrorist organization steal or otherwise acquire an operational nuclear weapons, and has this already happened? The short answer is that there are no confirmed reports, or even informed speculation, of nuclear weapons being missing or stolen from the stockpiles of the United States, Russia, the United Kingdom, France or China (or indeed India, Pakistan, Israel or North Korea). That does not imply that such an event could not, or has not, occurred. Neither should it be taken as a sign that theft of nuclear weapons from any of the nine states currently in possession of them is impossible.

Russia's nuclear security difficulties since the end of the Soviet era provide the bellwether in this regard, but it is not alone. With a great deal of American assistance via the National Nuclear Security Administration and the Department of Defense, Russia has dramatically improved the security of its nuclear stockpile, although serious problems remain. These range from inadequate surveillance and physical barriers, to corruption and "insider" theft by disaffected and poorly paid technical and security personnel. This leaves most Russian nuclear facilities extremely vulnerable to large scale and determined terrorist attack of the kind perpetrated in the Moscow theater siege of October, 2002. In that case, around fifty Chechen rebels took some 850 civilian hostages, holding them in central Moscow for over two days and proving their willingness to die for their mission, values and beliefs.

While physical nuclear security is broadly improving in Russia, and in any case is much stronger in the US, UK, France and Israel (little information is available on China) there are grounds for considerable doubt in regard to Pakistan, and to a lesser extent India. A recent nuclear security assessment notes that, while Pakistan appears to have adequate physical security measures in place, the threat to its nuclear weapons stockpile is centered on the combined threat of "insider" subversion and external terrorist attack. With the al-Qaeda terrorist network now based within its borders, and facing serious social unrest and possible economic collapse, Pakistan is a prime target for extremist terrorist attack on a very large scale. Pakistan has responded by enhancing its nuclear command, control and security systems, but attempted theft of its nuclear weapons by terrorist groups, on the Moscow scale noted earlier, must remain a live issue.

Finally, the possibility of a Pakistani nuclear "insider" group diverting a nuclear weapon for terrorist use must be considered plausible in light of the nuclear black market activities of A.Q. Khan. Khan, once at the summit of Pakistan's nuclear program, systematically engaged in the corruption of his own security structure to allow his organization to offer for sale both advanced centrifuge technology and weaponized nuclear explosives designs. Although he was eventually stopped, Khan's egregious activities serve to underline the entrenched and systemic nature of official corruption within Pakistan, often supported and sustained by senior military personnel with extreme political views. Should such individuals or small groups be charged with the security of nuclear weapons or their components, the risk of nuclear weapons theft becomes dangerously real.[19]

Covert International Trade in Nuclear Materials and Technologies

Nuclear fissile material, and other nuclear materials and technologies, are the subject of a large and legitimate international trading system, undertaken in support of the global "bargain" struck by the Nuclear Non-proliferation Treaty which came into force in 1970. The treaty represented and recognized the need for a compromise position in respect of the dual uses of nuclear technologies and materials – in civilian power production and the development of weapons of mass destruction. It continues to do so today, not having seen any alteration whatsoever to its terms in forty years. During that period, the global nuclear power generation industry has experienced mixed fortunes as the price of uranium on world markets has fluctuated. Nevertheless, the headline statistics on the scale of global nuclear power are impressive.

By early 2009, 438 nuclear power plants were operational across the global in thirty-one states with a total capacity of some 371 gigawatts-electric (Gw (e)), or around 16 per cent of the world's installed static electrical power. Of the thirty-one nuclear states, eight accounted for approximately 80 per cent of the total – the US, France, Japan, Germany, Russia, South Korea, Ukraine and Canada. An additional forty-four plants with an installed capacity of 39 Gw (e) were under construction in fourteen countries. Of these, the lion's share were in China (11), India (6) and Russia (8).

Given the scale of the global nuclear power industry, the size of legitimate world trade in nuclear materials and technologies involving transfers across international

19 Supra. At Note 5, pp. 32–5.

borders can be seen in context. Newly mined uranium is shipped around the world, enriched in its fissile isotope U-235, transferred for fabrication into fuel elements, sent elsewhere for use in nuclear power plants, often then reprocessed to form new mixed oxide uranium/plutonium fuel, or stored awaiting final deep, long-term sequestration.

Continuing Development of State-based Nuclear Weapons Programs

In all this, the potential for nuclear materials (as well as technologies) to be diverted by the governments of states for illegal use is obvious and great, and was increasingly recognized as such following India's 1974 "peaceful" first nuclear test, which exploited civilian nuclear materials and technology originally supplied in good faith by Canada. This event was the trigger for intensified efforts to control nuclear commerce, such as the Nuclear Suppliers Group (NSG) of countries engaged in international trade in dual use nuclear materials technologies – those needed for both power generation and nuclear weapons.

Unfortunately, as controls tightened over state-to-state nuclear transfers, so came the emergence of the private sector as a significant source for nuclear weapons proliferators. Pakistan is known to have exploited this growing avenue of access to nuclear weapons capacities over thirty years and more. Its success in doing so was the foundation for its ultimately successful nuclear weapons program. Other aspirant nuclear weapons proliferants followed Pakistan's lead – notably both Iraq and Iran, but similar goals were also evident in the activities of India, North Korea, Libya, Brazil, Argentina, Egypt, South Africa, Syria and Israel.

Over the years, this "gray" international market between nuclear aspirants increasingly became a black market in nuclear-related goods such as gas centrifuge uranium enrichment hardware, specialized metallurgy, specialized machine tools, and the myriad sub-components needed to construct an effective and reliable nuclear weapons program. It involved the illicit acquiescence and active cooperation of many commercial enterprises, more often than not from wealthy and technically proficient Western states such as Germany and the United Kingdom.[20]

The many initiatives resulting from this growing and dangerous phenomenon, from the 2004 United Nations Security Council Resolution 1540 (criminalizing the proliferation of weapons of mass destruction) to the recent US sponsored Proliferation Security Initiative, and much else, will be examined and assessed in subsequent chapters. Even so, the implications of a robust and resistant international trade in nuclear related equipment and materials of all kinds for nuclear terrorist operations are as obvious as they are, indeed, terrifying.

---

20 International Institute for Strategic Studies, *Nuclear black market dossier: A net assessment*, Chapter 2: "Nuclear Black Markets: Other Countries and Networks". Available online at: http://www.iiss.org/publications/strategic-dossiers/nbm/nuclear-black-market-dossier-a-net-assesment/. Accessed February 15, 2009.

This chapter serves to introduce the nature and scope of the growing security challenges posed by the rapid transformation evident in global nuclear affairs. The nuclear bargain expressed in the terms of the 1970 Nuclear Non-proliferation Treaty (NPT), an international legal instrument of almost universal scope, was negotiated during an era in which rampant nuclear weapons diffusion was seen as a real, and potentially catastrophic, danger.

For nearly forty years, the NPT and the regime of nuclear non-proliferation which grew around it, enjoyed moderate success as measured by the limited capacities of the nuclear proliferators to fulfill their aims. That success is now fading as the systemic flaws in global nuclear non-proliferation activities worldwide are found increasingly wanting when tested by determined and implacable actors of many kinds. The single disaffected security guard or technician who walks out of a nuclear facility with highly enriched uranium in his pocket; the cabal of senior Pakistani army officers with access to Pakistan's nuclear weapons stockpile; the al-Qaeda terrorist cell and its extremist collaborators in the building of a crude but operational nuclear explosive device; the national leader who sees his hold on power as dependent on his ability to construct and defend his country's nascent nuclear weapons program: all contribute in different ways to the seemingly inexorable rise in the level of nuclear threats of all kinds.

Thankfully, there is a complex matrix of feasible and effective responses to the growing threats to the homeland security of all states posed by these kinds of developments. However, before canvassing these more closely, it is essential to understand the international legal, political and technical institutional frameworks which form the context for global recognition of the nature of the nuclear threat, as well as prospective and substantial agreement on the way forward in response, and the longer term measures which may, eventually, put the nuclear genie – if not back in the bottle – at least under effective discipline and control.

Chapter 3
The International Institutional Framework

Introduction

The rising level of threat posed by illicit access to highly enriched uranium and plutonium, stolen or otherwise obtained by terrorist groups, or by others on their behalf, is at once highly complex in its structure, uncertain in its contingencies, and both broad and deep in its legal, political and technical contexts and reach. Simply put, it is no easy task to impose analytical coherence and accessibility on so complex a problem as nuclear fuel security.

In order to do just that, it is therefore essential that the undertaking be broken down into logically and structurally defensible elements. While this approach will be present throughout this book, it will only be effective in its goal of resolving how best to strike the most effective balance between alternative and complementary policy approaches to nuclear fuel security issues if it is undertaken within an appropriate institutional framework.

Taking an individual, domestic (or state), regional and global "levels of analysis" approach to building a vision of the nuclear institutional framework is a necessary first step. It acknowledges the reality that particular types of fissile material security threat (for example, the physical security of fully assembled nuclear weapons) will differ in terms of the nature of the threat, its perceived magnitude and extent, the breadth of appropriate strategies available to counter the threat, and the range of likely barriers to successful security enhancement. Any specific or discrete threat will, necessarily, be identifiable according to a range of influencing factors at the individual, state, regional (or alliance), and global levels.

Those factors are many, but may comprise combinations of, for example, the activities of individuals and non-state groups such as terrorists, their organizations and affiliates, multilateral nuclear security organizations and arrangements, the various national, regional and global security strategies of states (either singly or in alliance with others) and their domestic political constraints, as well as the technical sophistication and capacities of their economies, and of their domestic and global scientific and technical communities.

This suggests a second analytical approach to the imposition of coherence over complexity: distinguishing, building and then amalgamating three separate institutional frameworks, based on the legal, political and technical aspects of identified security threats. The two approaches, in conjunction, form a matrix within which most narrowly identifiable security threats can be defined and located, and one which can therefore contribute to the creation of a body of universally applicable response sets.

To illustrate this process, consider a hypothetical case in which a significant quantity of highly enriched uranium is sited in an inadequately secured university nuclear research laboratory located in a politically liberal unitary state. The state has an advanced economy and a globally acknowledged neutral foreign policy status, but is without a nuclear weapons program or civilian nuclear power industry. As a result, its moderate technical and scientific capacities to respond effectively to its growing nuclear security challenge limit its indigenous options in this direction. It thus acknowledges the degree of its dependence on other states with greater capabilities in terms of formulating and building its nuclear security options.

The HEU is held and used for legitimate scientific research purposes (in this case, medical isotope research) but nonetheless is increasingly regarded internationally as posing a significant and growing global nuclear risk in its vulnerability to attack and theft by terrorist or organized criminal networks.

The task here is to locate this highly specific nuclear fuel security threat within an institutional threat framework. In this case, the identified threat can be classified as applicable at the individual (terrorist) level, the nation state level, and at the global or universal level, since any successful terrorist-led theft of the HEU in question has security implications for the state in question, as well as for the entire world and its people. Furthermore, in an institutional sense this particular threat can also be classified in its [international] *legal*, [domestic and international] *political*, and [scientific] *technical* aspects, both domestic and global.

Although the nature and interaction of these institutional domains will be discussed in greater detail below, it is sufficient here to note that the hypothetical state in question will have need, at the very least, to consider the following factors in response to its home-grown nuclear fuel security threat scenario. First, it will be under general, and probably more specific, obligations in public international law in respect of the dangerously insecure HEU held in its university laboratory. This is likely to comprise a set of conventional and customary legal obligations established by the state's accession to multilateral United Nations and other treaties on nuclear security, and by relevant UN General Assembly and Security Council Resolutions.

Second, its international status as an acknowledged neutral country with no military alliance relationships or obligations will condition the approach of other states, either singly or cooperatively, towards strategies for understanding and countering the perceived terrorist threat posed by its enriched uranium holdings.

Finally, this imagined state's advanced scientific and technical capacities in non-nuclear areas give it a range of unilateral response options which would not necessarily be available to a state with a similar nuclear security problem, but with inferior technical capabilities. However, its acknowledgement of its specialized nuclear technical dependence on external regional or global resources for assistance has been echoed, in reality, in the US-led programs to remove HEU from research laboratories in many countries around the world.

These two complementary analytical approaches to understanding the general scope and nature of nuclear security outlined above – a "levels of analysis" approach which is combined with a practical taxonomy of the nature of specific threats, and then merged to form a threat matrix – is the fundamental theoretical basis for the widely acknowledged 'regime' of global nuclear non-proliferation activities. This regime encompasses all aspects of resistance within the international community of states against the illicit diffusion and use of nuclear materials and technologies. It encompasses relevant international law, other bilateral and multilateral agreements, understandings and cooperative activities of many kinds in multiple issue areas, and technical cooperation across international borders, all in pursuit of stopping further proliferation of nuclear capacities directed beyond legitimate nuclear power generation.

The International Legal Framework

It is necessary here to state that a wide ranging survey of the origins, legal philosophy and developmental history of public international law, the Law of Nations, is beyond the scope of this discussion. However, several general observations can be made in order to set the international legal scene and establish the nature, status and utility of this unique legal system, with particular reference to nuclear affairs.

Importantly, public international law as it stands today is the product of a European international legal tradition dating back, ultimately, to the ancient city states of Greece. Its later development parallels the emergence of the modern state in Europe from the conclusion of the 1648 Peace of Westphalia which ended the devastating Thirty Years War in Europe between the catholic Holy Roman Empire and the German Protestant princes. As it developed further, notions of Natural Law and ultimately of fairness or equity became embedded in the nascent concept of international law, and remain today as one of its central organizing ideas.

International law, however, emerged as a discipline and instrument of diplomacy in the modern sense only in the latter part of the nineteenth century, gaining added momentum from 1919 following the end of the First World War in the fields of humanitarianism and the laws of war. This was an attempt, in concert with the new social scientific discipline of International Relations, to understand how that terrible and unnecessary conflict had occurred, and how international law and the study of relations between states could, by limiting the nature and legitimate circumstances of the use of state aggression, together help the world to avoid a repetition.[1]

The first point to note about public international law is that it remains a paradox. On one hand, its legitimacy and capacity for effective deployment remains contested

1 For a clear discussion of the origins and developmental history of international law, see: Koskenniemi, M. (2004) The history of international law today, *Rechtsgeschicht.* Available online at: http://www.helsinki.fi/eci/Publications/MHistory.pdf. Accessed February 23, 2008.

ground in many quarters, including, at times, the highest echelons of governments. On the other hand, however, public international law has now reached a point where many aspects of its focus, application and effects are welcomed around the world as providing the only universally acknowledged and legitimate *basis* for effective and cooperative problem solving by states.

It should come as no surprise to the reader that the former view is held most strongly by those whose interests lie within the ambit of the unfettered use of political and military power by the sovereign states. It is also keenly felt by individuals and groups within government and the academy who regard the sovereignty of states within the international political system as absolute. Since the states claim, and will defend against all comers, their sovereign independence – the very essence of their identity and existence – it follows, say such advocates, that no hindrance to that goal can be acknowledged or tolerated.

Furthermore, they often see little or no irony in the fact that the sovereignty they seek so jealously to defend is itself an international legal concept now at least four hundred years old. Notice here that the claim by states to independence of decision and action no longer extends to the waging of aggressive war. Chapter VII of the United Nations Charter, the 'Basic Law' of world affairs since 1945, outlaws the crime of aggression as a direct result of the tragedy of the Second World War. In its place is the capacity of states, either singly or in coalition, lawfully to defend themselves against attack as a primary principle of international law, and one which echoes many domestic legal systems, including English Common Law, in this respect. The United Nations Charter is thus an example, as an international covenant or treaty which parallels an established domestic legal principle, of two of the four sources of international law, the other two being customary law, and legal scholarship.[2]

The corollary of international law used as a sub-text for the use of force by states becomes clear when we consider the broad and growing fields of international law in the public domain. Even if we put to one side the great legacy of customary, and now treaty-based, humanitarian law protecting human societies and individuals during times of war (codified principally in the four Geneva Conventions of 1949) much else remains. For example, one of the true success stories of public international law in recent times has been the 1982 Convention on the Law of the Sea (LOSC), which lays down universal rules relevant to all uses of the oceans and their resources. Many years in the making, the LOSC stands as affirmation that international law is capable of bestowing enormous benefits on the international system of states, and thus on the world's people – and especially so when the advantages to be gained through cooperative action are clear and well defined.

Going further, there are now few arenas of international life which are not the subject of international law rules and scrutiny, and from which enormous benefits have not accrued. From civil aviation to environmental conservation, from the

2 The four sources of international law are outlined in Article 38 of the Statute of the International Court of Justice.

panoply of human rights claims to international migration and refugee law, as well as international trade and commerce, all have seen massive advantage effected through the universal acceptance of international law as a globally legitimate set of coordinated and complementary rules and norms which transcend the international boundaries and interests of states. The most telling aspect of the strength of this enormous global rule set is that the states have cooperated, over many years and in many ways, to create it. International law's reach continues to grow in this way, and endures because its application is universally expected and accepted.

In a palpable way, therefore, international law is important because the states, the building blocks of the international system, *say it is,* most clearly through their decisions and actions. Again, and in a very real sense, international law is subject to continual reinforcement, reinvention, invigoration and adaptation in order to maintain its relevance and preserve it as an international regime with the legal legitimacy which no other domain can offer. Although (save for institutions such as the International Court of Justice, the International Criminal Court and the various Ad Hoc Tribunals) it is largely self-enforcing, international law endures despite its shortcomings because there is no alternative system of universally legitimate negotiation and cooperation establishing the rights and obligations of its subjects *as against each other*. Furthermore, there is no other non-military mechanism which is sufficiently stable, durable and comprehensively accepted by states for the peaceful settlement of disputes between them.

The International Nuclear Law Framework

Having noted the general shape of the international legal framework, it is now a relatively simple matter to establish the shape and scope of nuclear international law, and to place nuclear fuel and fissile material security law within it. It is convenient to begin with the central pillar of international nuclear law, the 1970 Nuclear Non-proliferation Treaty (NPT), the single most important legal and political document in the history of the nuclear age, and one with almost universal state accession.[3] As the *sine qua non* for all global non-proliferation efforts, the NPT forms the legal foundation for the three cornerstones of non-proliferation, namely:

- prohibition against the proliferation of nuclear weapons, materials and technologies,
- comprehensive and permanent nuclear disarmament, and
- enhancement and promotion of the peaceful use of nuclear power.

---

3 Only India, Pakistan, Israel and North Korea stand outside the NPT as independent nation states. The full text of the NPT can be seen at Appendix A.

However, it is much more than merely a multilateral legal instrument, vital as it is in that global role. In a political sense, the NPT represents a bargain or accommodation between those states in possession of nuclear armaments in 1967 – the five acknowledged Nuclear Weapon States (NWS) – and those without them. In essence, the NPT as negotiated forty years ago reflected the readiness of the Non-Nuclear Weapon States (NNWS) – the other bloc of states parties to the Treaty, all of which did not then possess nuclear weapons – to agree not to acquire them.

For their part, the five NWS states agreed to rid themselves of their own weapon stockpiles, under "strict and effective international control" and in good faith, but not within any specified time frame. At the same time, the NNWS states would gain clear and legitimated access to, and use of, nuclear materials and technologies only in verifiably peaceful nuclear power programs, and for other peaceful purposes.

From this beginning, negotiated by the United States and the Soviet Union at the United Nations in Geneva during 1967–68, a regime of ancillary treaties and other agreements gradually developed. One element was represented by the International Atomic Energy Agency (IAEA), a United Nations affiliated agency based in Vienna, Austria which undertakes inspection and verification work throughout the world to ensure that the legally binding Safeguards Agreements it negotiates with the NNWS states are adhered to. Importantly, it has sole legal responsibility within the United Nations system for ensuring that nuclear materials used either for indigenous power generation or otherwise legitimately acquired, are not diverted into clandestine and illegal applications involving nuclear weapons or terrorist-led nuclear enterprises.

Again, the NPT regime's three complementary nuclear materials and technology export control organizations, the Nuclear Suppliers Group, the Zangger Committee and the Wassenaar Arrangement, are semi-formal groupings of nuclear exporting countries which self-impose coordinated export controls based on Trigger Lists of specific items of nuclear material, equipment and technology, as well as those with identified dual-use (both civilian and weapons related) nuclear applications. The work of these complimentary export control groups is distilled in a range of international treaties in respect of nuclear materials and technology, all of which fall within the remit of the IAEA:

- their safe movement across international boundaries and at sea,
- prohibition of their possession within geographical boundaries (such as the various Nuclear Free Zones),
- the prevention of marine pollution by dumping; partial and comprehensive nuclear testing; and future prohibition against the production of fissile materials, and
- the suppression of acts of nuclear terrorism. The IAEA and the three nuclear export control groups thus form important arms of the nuclear non-proliferation regime.

However, and returning to the international nuclear law framework, it is clear that much progress over the past two decades has resulted in a strong and growing list of legal measures to combat all forms of nuclear proliferation, including the theft and use of enriched uranium and plutonium by terrorist groups and their affiliates. Within the complexity of international nuclear law, a representative sampling of that relevant to the security of nuclear fissile materials and nuclear terrorism includes the following important measures:

- Convention on Nuclear Safety (1994)
- Convention on the Physical Protection of Nuclear Materials and Nuclear Facilities (1979)
- IAEA Code of Conduct for the Safety and Security of Radioactive Sources (2004)
- Vienna Convention on Civil Liability for Nuclear Damage (1997)
- UN Security Council Resolution 1540 (2004)[4]
- Joint Convention on the Safety of Spent Fuel Management and on the Safety of Radioactive Waste Management (1997)
- International Convention for the Suppression of Acts of Nuclear Terrorism (2005)

In addition to international legal instruments such as these, the International Atomic Energy Agency provides a comprehensive suite of codes of conduct, discussion forums, data bases, reports, technical expertise, inspection and verification services, and nuclear policy analysis. It makes these services available to its member states, as well as presenting many of them in the public domain.

Furthermore, nuclear security enforcement measures are now strengthening as a result of their closer international coordination and the level of national and international resources devoted to them. One signal example is the 2003 introduction by the United States of the Proliferation Security Initiative, an informal grouping of states with serious concerns in respect of the growing threats posed by nuclear proliferation world wide, and a determination to contribute towards its elimination. In the case of the PSI, this takes the practical form of using national military and para-military assets to interdict international shipments, by any means of transportation, which are suspected of involvement in illicit trafficking in nuclear materials and technologies. The PSI is now supported by over eighty states, and is a prime example of the use of informal arrangements, rather than strictly delineated and obligating undertakings, as a diplomatically easier way to advance the cause of nuclear non-proliferation. The PSI has thus far enjoyed moderate but useful success,

4 UN Security Council Resolution 1540 unanimously resolved that "... all States shall refrain from providing any form of support to non-state actors" attempting to obtain or use weapons of mass destruction, "adopt and enforce appropriate effective laws" to that effect, and "take and enforce effective measures to prevent the proliferation of Weapons of Mass Destruction".

including the interdiction of at least eleven suspect international shipments. It has also been criticized on the ground that its maritime activities may be in breach of the International Convention on the Law of the Sea.

It has been argued above that the presence of a broad sweep of international legal instruments provides an indispensable element of the policy matrix which is at the heart of the nuclear non-proliferation regime. While this is certainly true, it is also indisputable that in the world of sovereign nation states, an international legal system can only operate effectively if allowed to do so by the states themselves. Thus, we now move to consider the international legal framework of nuclear affairs from the viewpoint of the governments of states and their policy advisers.

International Law and Nuclear Fuel Security

Law as a conceptual tool and instrument of social regulation is, by its very nature, an inexact means of effecting a desired outcome. It is also slow and uncertain in its reaction to changing trends and circumstances, and to emerging legal and practical needs. The Law of Nations, the crudest and most ill-formed of all legal systems, is no different from any other (such as English Common Law) in this respect. This is partially a function of its origins, structures and approach to issues of interpretation and application, and partially an inevitable result of the need for widespread agreement among the nations states before changes with universal force can be put into effect. The failure of all attempts by states, which are usually made during the five yearly NPT Review Conferences, to correct the deficiencies inherent in the text of the Nuclear Proliferation Treaty, possibly the single most important multilateral treaty of all, is testament to this truism.

Nevertheless, the universal acceptance of the legitimacy of international law should not be discounted in its role as the conceptual foundation for some of the most distinctive and promising recent proposals for nuclear fuel threat mitigation. One group of proposals stands out in its potential effectiveness in denying HEU or plutonium to both proliferators and nuclear terrorists, while enhancing access to nuclear fuels for bona fide purposes. These proposals can be grouped under the title of "fuel banks" and are gaining ground as their benefits become more widely appreciated.

One example of a fuel bank was proposed in 2006 by the Nuclear Threat Initiative (NTI), in which nuclear fuel supplies to customer states in full compliance with their international legal obligations would be guaranteed by the International Atomic Energy Agency. This would amount to establishing a "supplier of last resort" service to states otherwise denied, for whatever reason, access to nuclear fuel vital to the stability of their domestic energy supplies and economic futures.[5]

5 Nuclear Threat Initiative, Press Release: "NTI/IAEA fuel bank hits $100 million milestone;" Available online at: http://www.nti.org/c_press/statement_Nunn_EU_fuel_bank_121008.pdf. Accessed March 23, 2009.

The Nuclear Threat Initiative, an independent and US-based non-profit policy advocacy organization, has argued that it is essential in terms of long-term global nuclear energy security that states be persuaded to relinquish any aspiration to embark on their own uranium enrichment programs.

However, like many nuclear arms control and nonproliferation advocates, the NTI acknowledges that they are likely to do so only if it is clear that there is a guaranteed alternative source of nuclear fuel, beyond the global marketplace, which is non-discriminatory, non-political and available to any state in full compliance with its obligations under the NPT. A fuel bank controlled by the IAEA has, furthermore, the best chance of fulfilling this requirement, according to NTI, while it will also have the effect of strengthening the international market by enhancing both supply of, and demand for, nuclear fuel.

In this fuel bank example, it is clear that a nuclear fuel supply guarantee proposal of this kind, backed as it would be by both the legal and political standing and acknowledged technical capacities of the International Atomic Energy Agency, has at least a reasonable chance of succeeding in its aims. Other proposals, such as the United States' 2006 Global Nuclear Energy Partnership (GNEP)[6] go further in their ambition to allow peaceful nuclear energy applications to expand while not increasing nuclear security threat levels. Although the GNEP proposal addresses issues involved with the use of new and advanced nuclear technologies to counter nuclear fuel threats (such as ceasing to separate plutonium from spent fuel during reprocessing), here too participation by states is sought on the basis that no state will be required to abandon any rights it currently holds in respect of nuclear-relevant international law. Without the imprimatur of the IAEA, the GNEP proponent states, taken together, may have a harder task than their IAEA fuel bank counterparts (largely comprising the United States and the European Union states) in convincing aspiring nuclear energy states of that fact.

In all of this, it must be stated that the political and technical aspects of the nuclear fuel security policy context are not, nor are likely to be, overshadowed by their legally based counterparts. As in many aspects of the international life of the world, international law matters because the states make it so. Nevertheless, in issue areas as important and potentially dangerous as nuclear fuel security, the stabilizing, legitimating and determinative roles of public international law can only become ever more clear and irrefutable.

The International Nuclear Security Policy Framework

Using the analytical matrix developed earlier in this chapter, the international nuclear security policy framework can be understood in its national (or state), regional and global dimensions, as well as its legal, political and technical

6 Global Nuclear Energy Partnership Home Page. Available online at: http://www.gneppartnership.org/. Accessed March 23, 2009.

characteristics. Domestic and international nuclear policy frameworks are necessarily distinct, but in the globalized security world of 2010 and beyond, no state can expect to establish and develop its own nuclear policy settings in isolation from (at the very least) those of its regional neighbors, its military allies, its technical collaborators, and often its major trading partners. As a result, it is increasingly true that a country's national goal of strengthening the security of its nuclear fuels and other materials will be formed with the international context and consequences of such a policy firmly in view.

Furthermore, initiatives in this direction at the international level will command the most widespread attention, and promise the best chance of success, when they are instigated and supported by the wealthy, economically and technically developed states – and especially when the proponent state is armed with nuclear weapons or possesses a mature nuclear power industry. Nonetheless, international nuclear policy is ultimately framed at the truly global level, where the almost universal accession of states to the principles established by the 1970 Nuclear Non-proliferation Treaty is most clearly apparent. Thus, it is essential here to build an understanding of the effectiveness of the various elements of the nuclear non-proliferation regime, including its complexities and contingencies, in responding to the growing menace of nuclear fuel threats.

At the risk of repetition, it is perhaps useful here to pause and restate the most fundamental reason for the primary focus of this book. In 2002 the US National Research Council warned that "crude HEU weapons [could] be fabricated without state assistance" and that "the primary impediment that prevents countries or technically competent terrorist groups from developing nuclear weapons is the availability of [nuclear material], especially HEU".[7] It follows that the only potentially effective means of denying access to operational nuclear devices (or the capacity to construct them) for nuclear proliferators and terrorists is to deny them access to both highly enriched uranium, and plutonium in all its isotopic forms. This goal is the single greatest and, not surprisingly, most difficult task confronting the nuclear security policy community.

As noted earlier in this chapter, the 1970 Nuclear Non-proliferation Treaty (NPT) forms the kernel of a much more complex and diverse regime of international law, and governmental and non-governmental organizations, understandings and goals which echo the principle aims of the NPT itself. To reiterate, the non-proliferation regime's *raisons d'être,* in their broadest terms, encompass prohibition against the proliferation of nuclear weapons, materials and technologies; comprehensive and permanent nuclear disarmament; and enhancement and promotion of the peaceful use of nuclear power. Over a span of some four decades the NPT regime's record has, to be generous, been patchy at best. Its many critics, such as the governments of India over many years, point to the inequity of the non-proliferation regime's institutional creation of durable nuclear "have" and "have

7 Supra. At Note 5: "Why highly enriched uranium is a threat". Available online at: http://www.nti.org/db/heu/index.html. Accessed March 23, 2009.

not" states – those legitimately with and without nuclear weapons. They decry the manifest failure of the Nuclear Weapon States (the United States, Russia, the United Kingdom, France and China) to divest themselves of all of their nuclear weapon holdings over a period of almost forty years. Furthermore, they note the reality of the accelerating global spread of nuclear weapons knowledge, capacities and materials to the point where the original five Nuclear Weapon States have now become, at least de facto, nine with the addition of India, Pakistan, Israel and North Korea. Finally, such critics denounce the failure of the NWS states to allow rigorous and complete inspection by the IAEA of their nuclear facilities. This, they allege, makes inevitable their clandestine use of civilian nuclear power assets for military applications.

On the face of things, then, the NPT and its surrounding regime has been at best only a qualified success. On one hand, four new nuclear armed states in four decades seems unequivocally far better than President Kennedy's pessimistic 1962 prediction of between fifteen and twenty-five nuclear states by 1975. Certainly, the end of the Cold War around 1989–90 generated heated discussion of the threat of a sudden and dramatic rise in the pace of nuclear weapons proliferation as the constancy of that era rapidly gave way to global strategic instability.

On the other hand, the deficiencies inherent in the NPT regime have been described by Jacques Hymans as its "... ambiguous and erratically enforced rules, myriad technical loopholes, and chronic underfunding...".[8] Even a more sanguine assessment than this could be justifiably expressed in terms of intermittent success, frequent and often dangerous set-backs, lack of flexibility and unresponsiveness to new challenges, the whole accompanied by varying degrees of mutual suspicion and gradual systemic breakdown.

To be sure, the NPT regime is much more than merely a international legal concept whose use is limited to the arcane world of the Law of Nations. It is, of course, also a creature of international diplomacy and strategy, and of the world of national interests expressed as commitments to protect and defend the common precept of the prohibition against armed aggression. In pursuit of that aim, states continue to attend to their alliances and coalition arrangements with other states in the context of the broader goals of international arms control, and certainly of the eventual elimination of nuclear armaments and capabilities, of all kinds, everywhere.

Nevertheless, lofty goals expressed in these terms must and are underpinned by a second regime (or accepted rule-set) which is broader than the NPT regime, but just as vital to the continued success of both: international law or the Law of Nations. The NPT regime without the imprimatur and foundation stone of law would stand on very shaky ground indeed. Likewise, the most important element of international law focus and application, that of international disputation and

8 Hymans, J.E.C. Think again: Nuclear proliferation, *Foreign Policy*, November 2005. Available online at: http://www.foreignpolicy.com/story/cms.php?story_id=3304. Accessed April 1, 2009.

conflict, would not survive a global age in which its status, principles and processes had been subverted by a headlong rush to nuclear arms. In that scenario, the dogs of war, once slipped, could not be recalled.

However, it is clear that President Kennedy's initial pessimism, as the United States and the Soviet Union continued their atmospheric nuclear weapons testing apace, was misplaced (though understandably so). The extent to which international law has acted as a brake on the propensity of nation states to seek nuclear protection or aggressive potential by building nuclear weapons will always be unclear within the complexity of nationally generated motive, strategy and decision. This is so today as it was in 1964. Nevertheless, what is difficult to plausibly dispute is the effect which the presence of international legal principles and precepts can have upon politically conceived, negotiated and implemented initiatives.

Consider the universal nature of international law and the degree of its acceptance across the full spectrum of the activities of the nation states and the lives of their people. With very few exceptions, states recognize the utility and inevitability of the regulating effects of international law, especially (ironically enough) when its application is likely to cause difficulties and dangers. The plain fact remains that a state's recalcitrance *now,* on *this* legal issue, is very likely to be widely remembered in *future* times when vital measures of assistance are denied it on grounds of simple disbelief based on past failings, or mere retaliation.

If for no other reason, then, caution of this kind in the face of international legal realities will often lead to desirable levels of cooperation between states in order that they at least be *seen* to be willing to contribute to the common weal. In that aim, international legal considerations have for many years taken an unobtrusive but essential role. One clear example lies in the various proposals made over recent years to quarantine the enrichment of uranium from potential weapons proliferators by establishing a globalized setting in which legitimate enrichment (largely for use in power reactors) is guaranteed to states, while illegitimate enrichment is denied. The former goal is to be applauded, with caution, for its vaulting ambition, while the latter is probably unattainable, either with or without the strength of current international law.

The international security policy framework is a legally based supra-national structure which provides a degree of stability, durability, universality and transparency unavailable from any purely political or diplomatic alternative. It establishes the limits of acceptable behavior in all aspects of the use of nuclear materials in any and each application, including the threats posed by deficiencies in the security of nuclear fuels.

The remainder of this chapter is focused on an examination of the effectiveness of the nuclear non-proliferation regime, both generally and in specified cases, in creating and maintaining the quarantine of nuclear fuels from agents of all kinds which would seek to use them for malign applications. The roles, successes, failures and systemic deficiencies which are becoming increasingly clear in the operation of the NPT regime will figure prominently in this exercise.

The NPT Regime and Nuclear Fuel Security: Success, Failure, Deficiencies and Future

Conventional wisdom, confronted by the role call of success *versus* failure on the part of the international regime of the Nuclear Non-proliferation Treaty in enhancing nuclear fuel security, is both necessarily circumspect and advisably pragmatic. The plain fact is that Hyman's criticisms of the NPT and its regime noted earlier are unavoidable in such complex artifacts of international diplomacy as these, fashioned as they are by flawed (albeit extremely well-informed and highly talented) human beings. The same is true of the specific issues addressed in denying nuclear fuels to terrorist organizations and state proliferators. As a result, it is as difficult to claim unalloyed success as it is unmitigated failure. The history of nuclear fuel security supports a position far more common in any continuum: a complex and mixed result lying somewhere between these extremes.

The suite of goals which are central in efforts to deny nuclear fuels to all agents without a legitimate reason for possessing them is well captured, in both direct and indirect ways, in Matthew Bunn's report for the Nuclear Threat Initiative entitled "Securing the Bomb 2008":

> **Secure:** Every nuclear weapon and every significant stock of potential nuclear bomb material worldwide must be accounted for and secured, in ways that will last, to standards sufficient to defeat the threats that terrorists and criminals have shown they can pose.
>
> **Remove:** Potential nuclear bomb material should be removed entirely from the world's most vulnerable, difficult-to-defend sites, and the total number of buildings and bunkers worldwide where nuclear weapons or nuclear bomb material exists should be cut dramatically.
>
> **Disrupt:** Counter-terrorist measures focused on detecting and disrupting those groups with the skills and ambitions to attempt nuclear terrorism should be greatly strengthened, and new steps should be taken to make recruiting nuclear experts more difficult (including addressing some of the sources of radical Islamic violence and hatred, and challenging the moral legitimacy of mass-casualty terror within the Islamic community).
>
> **Interdict:** A broad system of measures to detect and disrupt nuclear smuggling and terrorist nuclear bomb acquisition efforts should be put in place, including not only radiation detectors but also expanded international police and intelligence cooperation, increased emphasis on intelligence operations such as "stings" (that is, intelligence agents posing as buyers or sellers of nuclear material or nuclear expertise), and targeted efforts to encourage participants in such conspiracies to blow the whistle.

Prevent: The nuclear programs in North Korea and Iran must be capped or rolled back, and the global effort to stem the spread of nuclear weapons should be significantly strengthened, reducing the chances that a state might provide nuclear materials to terrorists.

Deter: The United States should put in place the best practicable means for identifying the source of any nuclear attack—including not just nuclear forensics but traditional intelligence means as well—and make very clear that the United States will treat any terrorist nuclear attack using material knowingly provided by a state as an attack by that state, and will respond accordingly.

Impede: The United States and other countries should leave no stone unturned to make it more difficult for terrorists to recruit experts who could help them with a nuclear plot, from providing alternative employment for key experts to countering the anti-American hatred that can contribute to such recruitment efforts.

Reduce: The global stockpiles of nuclear weapons, separated plutonium, and HEU should be drastically reduced, and new production phased out.

Monitor: Stockpiles of nuclear weapons, separated plutonium, and HEU—and reductions in these stockpiles—should increasingly come under international monitoring, facilitating cooperation in upgrading security and ensuring a baseline level of accounting for these dangerous arsenals.[9]

It is beyond the scope of this chapter to comprehensively canvas the vast landscape of the NPT regime, its individual successes and failures and the lessons they teach, even in respect of fuel security alone. This latter task will be undertaken in Chapter 5. Instead, the performance of a handful of its prominent constituent parts, such as the Global Initiative to Combat Nuclear Terrorism, the Proliferation Security Initiative, and the G8 Global Partnership Against the Spread of Weapons and Materials of Mass Destruction, will serve as analogs for wider truths about fuel security within the international framework of the nuclear non-proliferation regime.

The three measures noted in the previous paragraph are touchstones for the very nature of the NPT regime, much of which can be characterized as ad hoc and reactive, rather than coordinated and proactive; partial, rather than overarching; and driven more by independently conceived notions within states' elite classes of national security and interests, rather than by an integrated conception of the nature of the global nuclear threat.

---

9 Bunn, Matthew. *Securing the bomb 2008*. Cambridge, Mass. and Washington, DC: Project on Managing the Atom, Belfer Center for Science and International Affairs, Harvard Kennedy School and Nuclear Threat Initiative, November 18, 2008. At pp. 1–2.

Over some forty years, attempts to crystallize and realize the vision of the negotiators of the NPT have been hedged about by innumerable impediments and barriers to progress, ranging from genuine fears of compromised national security outcomes (such as the enduring diffidence of the original five nuclear weapon states in addressing their own obligations towards complete nuclear disarmament) to disingenuous protestations on the necessity of nuclear weapons ownership based on perceived need to support and bolster a country's national prestige and regional security (in the case of India and Pakistan).

As a result, the NPT regime has developed largely without recourse to deliberative or carefully integrated planning and coordination of negotiation, diplomacy or action. For example, an initiative by the Board of Governors of the International Atomic Energy Agency on, say, issues involved with its notification to state authorities of imminent nuclear facilities inspections, while important, is likely to go largely unreported and barely noticed beyond those governments and their agencies directly affected by the changed inspection and verification conditions.

Meanwhile, the governments of European middle power allies of the United States fail to install adequate systems of oversight and controls of their exports to Iran (trans-shipped through legitimate third party countries) of highly specialized equipment and materials with uranium enrichment and nuclear reactor operations applications. In an egregious example of attempts by commercial enterprises to circumvent national export restrictions on dual-use technologies, some 50 German companies were prosecuted in 2007, in Potsdam, Germany, for exporting such goods to Iran, via Russia and Dubai, for use in the Russian-built and nearly complete Iranian light water reactor at Bushehr.[10] What is unclear is the reason for the failure of the German export regulatory authorities to identify, target and prevent this situation from developing. The same questions would apply, with even greater moment, had the export shipments involved been highly enriched uranium or plutonium charged fuel rods.

Consider the following three examples of attempts to rein in threat levels embedded in the legitimate use of nuclear fuels.

The Global Initiative to Combat Nuclear Terrorism

On its face, the Global Initiative to Combat Nuclear Terrorism appears to find a good fit within the panoply of international non-proliferation activities. It has been noted by Bernstein as one example of the "partnership imperative" initiated by the United States as a central US strategy to combat the proliferation of weapons of mass destruction through practical cooperation, better compliance enforcement,

10 Update: German and UK investigations target nuclear exports to Iran, *WMD Insights*, September, 2007. Available online at: http://www.wmdinsights.com/I18/I18_EU3_UpdateGermanUK.htm. Accessed April 6, 2009.

and the building of enhanced trans-national defense and response capacities.[11] The partnerships created are designed to be informal in nature, and capable of attracting large numbers of like-minded states, whose responsibility it is to take independent or coordinated sovereign action when the need is clear, and effective response is judged by them to be feasible.

In the case of the Global Initiative, early progress seems to have been accomplished. This program was established by the United States and Russia in 2006 as a means of boosting practical cooperation between all states concerned with implementing the goals of recent international legal initiatives countering nuclear terrorism. In particular, the Initiative was guided by the thrust of the 2005 International Convention for the Suppression of Acts of Nuclear Terrorism, the 2005 Convention on the Physical Protection of Nuclear Materials, and the 2004 United Nations Security Council Resolution 1540 requiring and authorizing sovereign domestic measures to control the spread of weapons of mass destruction.

In essence, the Global Initiative to Combat Nuclear Terrorism is a capacity building enterprise. As such, it aims to enhance global capabilities to successfully account for, control, and physically protect nuclear and radioactive materials in civilian nuclear facilities; detect, search for, confiscate and control illicit nuclear materials; deny safe haven and financial resources to nuclear terrorists; support national legal and regulatory frameworks; respond to, mitigate and investigate cases of nuclear terrorist activity; and finally enhance information sharing between states.

What makes this international nuclear control measure distinctive in comparison to much of what has gone before in the field is the confidence of its proponent states that it is possible to balance the informality (and thus flexibility) of the diplomatic process and its practical implementation, with the benefits to be garnered from reinforcing its legal standing through the binding force of UNSC Resolution 1540 noted earlier. Taken together, this creates a potent and extremely effective means of drawing a line in the sand for nuclear terrorists and, of course, for homeland security. By 2008, 66 states had joined this informal initiative, which together had identified a large range of cooperative activities directed at its specific aims noted above. It should also be noted, however, that many activities, such as nuclear terrorism law enforcement training, and bilateral capacity-building exercises, have been undertaken by the United States alone, and largely on its own account.

As a nuclear fuel security enhancement measure, the Global Initiative to Combat Nuclear Terrorism is not merely a single, isolated example of its kind. Other initiatives have appeared and been developed in parallel, one such being the 2003 US-sponsored Proliferation Security Initiative.

11 Bernstein, P. I., International partnerships to combat weapons of mass destruction, *Center for the Study of Weapons of Mass Destruction Occasional Paper 6*, National Defense University Press, Washington, DC, May, 2008.

The Proliferation Security Initiative

Even more pragmatically driven than the Global Initiative to Combat Nuclear Terrorism, the Proliferation Security Initiative (PSI) is, once again, an informal grouping of states which is pledged to act within the constraints of international and domestic laws, but which operates beyond the purview of any existing multilateral export control regime (such as that of the Nuclear Suppliers Group). An initiative in 2003 of the Bush Administration, its operational doctrine is to allow informality and flexibility to transcend the rigidities and disputes of more formal arrangements. Illicit shipments of nuclear materials are liable to interception on the high seas by participant states, a possibility with serious implications for any state complicit in assisting with such transfers. To that end, special attention has been given to maritime shipments to and from suspect states such as North Korea, Libya and Iran.

In its first five years, membership of the PSI grew from its 11 founding states to number 80, with more joining, which represents a significant proportion of the international community of nation states. While appraisals of its effectiveness have varied somewhat, the PSI has successfully conducted over 25 interdiction exercises, and performed a number of nuclear materials interceptions. The most widely known is the October 2003 maritime interception of uranium centrifuge parts consigned from Malaysia and bound for Libya aboard a German-owned ship (Germany is itself a founding member state of the PSI).

Notwithstanding its successes, the PSI is the target of substantial criticism based on some degree of uncertainty about the legal status of its aims and activities, and the undeniable fact that it is dominated by the United States. As yet another unilateral and adjunctive measure within the broad ambit of the NPT regime, and one with nuclear fuel security and homeland security firmly in its sights, the PSI is vulnerable to criticism by those who see it as more clearly an American, and less clearly a global, nuclear security and counter-proliferation effort. Furthermore, there appear to be more than insignificant grounds for legitimate concerns, at least on some issues.

For example, it is accused of vagueness of purpose and secretiveness in its operations, and that its informality limits its effectiveness and sustainability. India has displayed resistance to the PSI as undermining its foreign policy independence; Malaysia and Indonesia worry about its international legality, and the intentions of the United States in respect of territorial sea claims, the US not having ratified the Law of the Sea Convention.[12] Another popular critique points out that a state or non-state agent wishing to avoid detection can do so by simply using their own flag vessels or aircraft, or those of non-participating states, which cannot be legally searched on the high seas. A further operational hiatus exists, say some critics, in the fact that targets for interdiction of nuclear materials and fuels are limited to commercial shipments only, thus allowing all government shipments to

12 Ibid. At p. 5.

escape scrutiny. Finally, there is strength in the argument that the PSI does not alter the fundamental dilemma surrounding all non-proliferation efforts: that it will not capture the many dual-use technologies which any state can employ in order to develop a covert nuclear weapons program under cover of a power generation program.[13] It may well be that PSI is simply incapable of effective use.

The G8 Global Partnership Against the Spread of Weapons and Materials of Mass Destruction

In contrast with the broadly based and somewhat inchoate Proliferation Security Initiative of 2006, the G8 Global Partnership Against the Spread of Weapons and Materials of Mass Destruction (G8-GP) has an altogether more potentially effective provenance: the resources of the eight wealthiest countries when coordinated and focused on putting nuclear security and non-proliferation ideas and aspirations into practical effect. An outcome of the 2002 Meeting of the G8 group of nations in Kananaskis, Canada, this program committed the G8 group of countries to raise US$20 billion by 2012 towards a range of non-proliferation projects, principally in Russia and Ukraine. Among them are programs to enhance security for the storage of spent HEU once used to fuel decommissioned Russian nuclear powered submarines, as well as for hundreds of metric tons of other Russian weapons-usable nuclear material.[14]

Building on previous cooperation between the United States and Russia during the nineteen nineties, the G8-GP program operates as a multiplier of effort through a number of distinct counter-proliferation activities funded through the G8 process. Prominent among them are those which target insecure and unwanted Russian HEU and plutonium holdings. The US continues to purchase and downblend some 500 metric tons of HEU from Russian warheads for commercial energy production within the United States, as well as working to help in the decommissioning of Russia's plutonium fueled production reactors and dispose of 34 tons of its surplus plutonium stocks. Finally, the G8-GP process is funding efforts to eliminate HEU in research reactors around the world, and to secure and eventually remove all fresh and irradiated HEU holdings.

The G8 Global Partnership program is, unsurprisingly, not without its critics. It is an ambitious and complex international initiative which has inevitably encountered obstacles to success, such as the preponderance of focus on Russia, to

13 Valencia, M.J., *Bring the Proliferation Security Initiative into the UN*, Policy forum online 05-101A, Nautilus Institute, December 20, 2005. Accessed April 12, 2009.

14 Supra. At Note 12, pp. 7, 8. The original G8 Global Partnership members comprised the United States, the United Kingdom, Russia, France, Japan, Italy, Canada and Germany. They were subsequently joined by 13 others: Australia, Belgium, Czech Republic, Denmark, Finland, Ireland, Netherlands, New Zealand, Norway, Poland, South Korea, Sweden and Switzerland.

the possible detriment of pressing security issues elsewhere. Another criticism is prompted by the dominance of the United States in efforts to upgrade security for Russia's nuclear materials stockpiles, leaving its partner states, allegedly, barely in the game.

Conclusion

As a limited group of international programs with significant influence and standing within the nuclear non-proliferation regime, these three initiatives – chosen more or less at random – are familiar for several of their larger features: origins founded in limited international strategic initiatives; support dominated by the United States, with less than proportionate contributions from other participant states; elements of ad hoc reactivity, rather than carefully coordinated and cooperative effort over time; and varying degrees of vagueness of guiding aims, combined with the prospect of duplication of effort across similar programs.

The conclusions suggested by these considerations are, prima facie, as true of the broad remit of these broadly based nuclear non-proliferation programs as they are in terms of the nuclear fuel security questions which are the specific focus of this discussion. As this chapter has suggested, the dual fields of international nuclear law and international nuclear security policy come together in all efforts to deny nuclear materials, technology, weapons-building capacities and ownership to state and non-state agents which threat homeland security for the United States and elsewhere. Where each field of endeavor is found, so too is the other.

The result is that programs and policies built to assist in the cause of enhancing nuclear fuel security on a global scale are far from incoherent, far from operating as unique exceptions to the universally accepted rules of international law and diplomacy. They are, in contrast, embedded within such a conceptual framework *just because* the threats posed by malign use of nuclear fuels are so great for the world and its people. Nevertheless, much is still to be done if even modest gains in the global coordination of nuclear non-proliferation effort and – by extension – nuclear fuel security enhancement are to be attained.

Where to from here on this vital and enormous question? It is worthwhile to introduce here the possibility of using the unique nature of the single worldwide clearing house for global nuclear security, the independently established and United Nations affiliated International Atomic Energy Agency, as the starting point in constructing the shape and future of coherent, efficient and effective action towards a more secure nuclear world. The development of the complexities of this proposal must await later discussion, but its feasibility is worthy, at the very least, of initial consideration by all those now watching the breakout of security threats inherent in the world's nuclear fuels.

Chapter 4
Assessing Security Threats Inherent in Nuclear Fuels

Introduction

The previous chapter introduced the international institutional framework of nuclear non-proliferation through an examination of several of its most crucial elements, such as nuclear international law and the international nuclear security policies of states, both singly and in alliance. From this foundation, and from earlier discussions on the nature and extent of global nuclear security threats thus far undertaken, it is now necessary to consolidate and characterize the full suite of security threats inherent in existing and projected stocks of highly enriched uranium and plutonium held around the world. This builds on the discussion of broad categories of threat relevant to nuclear terrorism undertaken in Chapter 2, and takes account of the quantities of nuclear fuel stocks held around the world, and noted there.

This exercise is necessary in order to establish the patterns of resource allocation and time scales within which a more comprehensive and better coordinated program of enhanced global nuclear fuel security measures must be undertaken. This will be discussed in later chapters but, with that aim in view, this chapter examines the major nuclear fuel threat elements in turn, in order to assess the relative urgency of response demanded by each. Denial of access to highly enriched uranium (and, to a lesser degree, plutonium) for use in an improvised nuclear device by terrorists, or in a covert nuclear weapon program by states, is the world's best chance to put effective and, most importantly, lasting nuclear non-proliferation strategies in place for the twenty first century and beyond.

As noted earlier, the security status of these two nuclear fuels holds the key to understanding and measuring the nature and levels of nuclear threats now facing the world's people. In addition, perceptions of the range of possible threat scenarios, and appropriate responses to them, is diverse across the society of states. This demands that the greatest care be taken by nuclear negotiators, policy planners and decision makers in avoiding unnecessary misunderstandings in the arena of nuclear fuel security policy development. It also requires that they be willing to work in more strategically and practically coordinated ways in countering this most serious of globally existential threats.

In this last context, a note is necessary here about the specific dangers of highly enriched uranium, as compared to plutonium, if stolen and used by non-state agents such as terrorists to build nuclear explosive devices. Simply put, it is much easier to put together an improvised nuclear device (IND) using enriched uranium than one fueled by plutonium. In the case of a uranium device, a simple gun-type design, using a suitable artillery barrel to fire a sub-critical projectile of metallic HEU at a sub-critical target of the same material, should result – under suitable conditions – in a super-critical chain reaction and nuclear explosion.

In contrast, the use of plutonium as the fissile element of an IND requires the employment of complex lens implosion and timing techniques in designing and building an operational bomb, and is therefore much more difficult to construct and detonate successfully.[1] The conclusion must therefore be that, at least in its initial phases, a renewed global nuclear fuel counter-proliferation effort should, all else being equal, concentrate first on securing, consolidating, reducing, and finally eliminating global holdings of HEU, before moving to further reduce, and then eliminate threats from free, and then unseparated, plutonium.

Having examined the nature of the threat of nuclear terrorism, the chapter considers the adequacy of international measures to safeguard the use of fissile materials in both the Nuclear Weapon States (as both defined in the Nuclear Non-proliferation Treaty of 1970, and including other nuclear-armed states) and states without nuclear weapons.

The chapter goes on to discuss the Pakistan-based clandestine nuclear technology trading network of A.Q. Khan, known colloquially as the father of Pakistan's nuclear weapons program. As noted earlier, Khan's network provides the most important and egregious example so far uncovered of the malign influence of senior and well-placed individuals and specialized organizations in non-state and quasi-state nuclear technology proliferation. This case, involving the alleged transfer of nuclear weapons technologies between Pakistan and Iran, Libya and North Korea, is emblematic of the threats posed generally by agents operating beyond the sphere of state authority while functioning within it, or with the tacit approval or acquiescence of governments. It is only a short step from offering for sale, to the highest bidder, uranium enrichment centrifuge components or kryton switches used for detonating a nuclear fission device, to placing HEU or plutonium on the international armaments black market. As the Khan network demonstrated until 2004, there is likely to be no shortage of customers for similar organizations today and in the future.

The chapter concludes with an introductory consideration of the resource and timing constraints which will increasingly be placed on nuclear fuel security planning as the boundaries of nuclear weapon proliferation continue to be challenged.

---

1 Of the six isotopes of plutonium, Pu239 is by far the most important here, being the most readily fissile at a rate suitable for producing a sustained nuclear chain reaction. Unless otherwise noted, data on the specific threat posed by HEU, in contrast to plutonium, is taken from: Potter, W.C., Nuclear terrorism and the global politics of civilian HEU elimination, *Nonproliferation Review* vol. 15, No. 2, July 2008.

Nuclear Fuels: What is at Risk?

Chapter 2, Table 2.1 contains broad estimates of global stocks of HEU and plutonium assigned to civil and military use. Over 3,700 metric tons (mt) of fissile material suitable for use as nuclear weapons is held in civil and military programs. Excluding unseparated plutonium in spent reactor fuel, the total of more readily usable fissile material of both kinds, held in varying and often inadequate or inappropriate conditions of physical security at some 350 sites in nearly sixty countries, is still around 2,300 mt, enough for more than 200,000 nuclear weapons.

Nuclear Fuel Threat Categories

The most coherent way in which to establish a field of nuclear threat categories is to divide them between threats relevant to the civilian and military domains. Within each of these, there are a number of distinct threats which can be further understood in terms of their immediacy, contingency and consequences; if threat becomes reality, how soon can it be expected, under what conditions, through what proximate and background causes, and with what results for individuals and for states? The appropriate response in each case will amount, of course, to a distillation of the questions: what must be done, what can be done, when, and with what results?

Civilian Fuel Threats

Nuclear fuel threats within the civilian domain can also be divided initially between two main categories of threat. The first is a set of threats involving the theft of, trade in, and use by terrorist organizations, of HEU or plutonium. The second is concerned with the diversion to weapons development programs of nuclear fuel held legitimately for use in civilian power generation programs, but suitable for use in nuclear weapons. Within the first category, activities perpetrated by, or on behalf of, extreme terrorist organizations must be clearly distinguished from state-led covert diversion of fuel from existing civilian power production programs, since the suite of available responses in each case is significantly different. However, the two categories are, of course, linked to the extent that fissile materials diverted by state agents are, *ipso facto*, thereby transformed into targets for use in nuclear terrorist attacks.

Nuclear Theft

It might be thought by those with a general understanding of the nature and uses of nuclear fuels that state security authorities and legitimate non-state agents such as commercial nuclear power generation corporations with responsibility to ensure its safe-keeping would, almost automatically, have long ago put in place and maintained the most stringent security regime possible to guard against theft. While this has been broadly so in the West for many years, it is far from true elsewhere. Theft of fully functional, metallic, weapons grade HEU must be regarded by aspiring nuclear terrorists as by far the easiest route to their goal, bypassing as it does the need to obtain uranium fuel precursors such as uranium oxide or uranium hexafluoride, which require extensive, difficult, lengthy and enormously costly processing and enrichment to convert them for weapons use. The same principles apply in respect of metallic weapons-grade plutonium, as compared to the plutonium present in, but not separated from, spent commercial reactor fuels, adding to the difficulties involved in devising an operative plutonium weapon (as described above).

Given the gulf between the utility of readily usable fuels and their alternatives, the deficiencies present in non-West fuel security measures, following the end of the Cold War in the early nineteen nineties, were both stark and extremely dangerous. Principally relevant to Russia as the successor state to the Soviet Union, but extending to several former Soviet republics such as Ukraine and Georgia, a vast complex of nuclear materials, stored in many poorly secured buildings and bunkers, came under increasingly severe security threat. While Russia has made substantial progress in addressing weapons-grade fuel security since that era, weaknesses remain in the form of storage site diffusion, poorly trained and under-paid guards, the possibility of insider theft, and inadequate technical security systems.

One recent example of an attack on a substantial nuclear facility in South Africa underlines the plain fact that theft of nuclear fuel suitable for weapons use is unfortunately far removed from fantasy. As reported by Matthew Bunn, South Africa's nuclear facility at Pelindaba, operated by the Nuclear Energy Corporation of South Africa, and where hundreds of kilograms of HEU are stored, was attacked by two teams of armed men on the night of November 8, 2007. One team fired on the Pelindaba security forces, who immediately fled, allowing the second team to disable detection systems and gain entry through a 10,000 volt electrified security fence. This group of gunmen gained access to the facility's emergency control center where they shot one worker dead, and were subsequently able to spend some 45 minutes within the facility's secure perimeter without being challenged. Both groups of gunmen then got away without injury.[2]

While the motives behind the Pelindaba attack are unclear, the lesson it provides for nuclear fuel security planners is not. For whatever reason, security at the site was inadequate, and significantly compromised. Even more concerning was the fact that the International Atomic Energy Agency's independent security assessment

---

2 Bunn, Matthew. *Securing the bomb 2008*. Cambridge, Mass. and Washington, DC: Project on Managing the Atom, Belfer Center for Science and International Affairs, Harvard Kennedy School and Nuclear Threat Initiative, November 18, 2008. At pp. 3–4.

regime, its Design Basis Threat assessment process for developing appropriate and effective facility protection systems, had been acknowledged by the South African nuclear regulator prior to the attack, but had not been implemented. In the event, no nuclear materials were stolen, but a similar attempt on a poorly secured Russian nuclear fuel repository, by a terrorist group of the size and potential present in the Beslan terrorist school massacre of 2004 in the North Caucasus region of the Russian Federation, may well have had a different outcome.

Trafficking in Nuclear Fuels

In 2007, the IAEA reported that, during the calendar years 2005 and 2006, some 250 incidents of unauthorized possession and other criminal activities, theft, disposal, and loss of radioactive materials had been detected and reported to the agency's Illicit Trafficking Data Base (ITDB). This outcome was described by the IAEA as "a significant problem". Of a wide range of specific activities, the most serious it noted were 14 cases of illegal possession, movement and attempts to trade in nuclear materials, mostly natural and depleted uranium, and thorium.

In January, 2007 Georgia reported to the ITDB the interception of a small quantity of 89 per cent enriched uranium. A further 85 incidents reported in 2006 involved the theft, loss or mis-routing of nuclear or radioactive materials, suggesting the presence of later trafficking attempts. Furthermore, in 73 per cent of reported case the material involved had not been reported as recovered. As of December 31, 2007, the ITDB contained 1,340 confirmed incidents recorded since 1993 by 100 participating states, and roughly divided between unauthorized possession and related criminal activities, theft or loss of material, and other unauthorized activities. [3]

Seeking to traffic nuclear fuels across international borders as supplier or customer, distinct from appropriation through outright theft, poses as serious a threat to the world's future as any other illicit element of nuclear proliferation. Both plutonium and low enriched uranium (LEU) are, of course, commonly moved within and between countries in the course of legitimate nuclear power generation activities. For example, up to 20 metric tonnes of plutonium are separated by reprocessing from spent nuclear power generating fuel rods annually in the United Kingdom, France, India, Japan and Russia. Much of this plutonium is then transported to fabrication plants for use in the production of mixed oxide (MOX) fuel, from where it is then moved to nuclear power generation plants. Again, although little or no HEU is now being produced for civil use around the world, much of the 175 tons of civil stocks of HEU held for research activities

3 UN News Centre, UN News Service, *Illicit trafficking, theft of nuclear materials "a persistent problem," UN Agency reports*, September 12, 2007. Available online at: http://www.un.org/apps/news/story.asp?NewsID=23774andCr=nuclearandCr1. Accessed May 5, 2009.

globally is progressively being transported long distances, for example for down blending in the United States.

None of these activities should, in any way, be compared to the illicit movement of HEU and plutonium across international borders and across the world, which certainly poses many active and growing nuclear security threats. Nevertheless, the legitimate movement of nuclear fuels also contributes to security challenges as they are carried within countries and across international boundaries by road, rail and maritime transportation.[4]

When considering the illicit trafficking of HEU and plutonium, the work of the International Atomic Energy Agency is a reliable measure of the practical difficulties encountered during counter-proliferation activities in interdicting nuclear fuels in transit. However, the capacities of the IAEA as a global nuclear watchdog are limited by the degree to which its member states are prepared to support it with financial, technical and political resources. Until recent times, the IAEA has struggled to maintain even its core verification and safeguards inspection and monitoring work, although the exposure of the Khan nuclear smuggling network has help to focus international attention on the range of non-proliferation regime deficiencies. The urgent and growing need to re-think, reform and renew all aspects of nuclear non-proliferation efforts is echoed in the words of the Secretary-General of the United Nation, Ban Ki-moon, in an address in May, 2009 in New York to the Preparatory Conference for the crucial 2010 NPT Review Conference. In it he stated:

> … nuclear weapons remain an apocalyptic threat. We cannot afford to place disarmament and non-proliferation on a backburner. Let us not be lulled into complacency. Let us not miss the opportunity to make societies safer and more prosperous.

The UN Secretary-General went on to note:

> ... a rising chorus demanding action on nuclear disarmament and non-proliferation. Concerns about nuclear terrorism, a new rush by some to possess nuclear arms, and renewed interest in nuclear power as an alternative to fossil fuels have only heightened the need for urgent action.[5]

---

4 For a detailed discussion of the transportation of weapons-grade plutonium in the commercial nuclear industry, see: Allbright, David, Shipments of weapons-usable plutonium in the commercial nuclear industry, *The Institute for Science and International Security,* January 3, 2007. Available online at: http://www.isis-online.org/ISISplutonium_shipments.pdf. Accessed May 2, 2009.

5 UN Secretary-General, Office of the Spokesperson, Latest Statements, May 4, 2009. Available online at: http://www.un.org/apps/sg/sgstats.asp?nid=3824. Accessed May 6, 2009.

Before moving on to consider the general implications for resource allocation of nuclear security policy planning in respect of nuclear fuels, it is appropriate to consider the rising incidence of nuclear theft and international trafficking in nuclear technologies and materials in the context of the threat potential of activities of this kind. The archetype is the international nuclear trafficking network of Abdul Qadeer Khan, widely regarded as the originator of Pakistan's nuclear weapons development program through his activities in providing his homeland with gas-centrifuge uranium enrichment technology.

The A.Q. Khan Network[6]

The history of the emergence, activities and ultimate suppression of the A.Q. Khan nuclear trafficking network is similar to other large scale and illegal international operations: murky, not fully understood, yet carrying with it immense and durable implications for global security in the precedents it set. This was nothing less than the successful development of an integrated and comprehensive international nuclear smuggling network, beyond the purview of government, which offered to the highest bidder whole gas centrifuge cascades for uranium enrichment on a "turn-key" basis, as well as production facilities and even designs for nuclear weapons.

It origins can be found in Pakistan's rivalry with India, and specifically in its military defeat by India in December 1971. That conflict on the Indian sub-continent, which entailed Pakistan's loss of East Pakistan, created the new state of Bangla Desh. As a result of his country's defeat, A.Q. Khan, then working in the nuclear industry in the Netherlands, stole from his employer and offered to his government Dutch centrifuge uranium enrichment technology. His offer was accepted, with the result that, within a decade, Pakistan had attained the capacity to build operational nuclear devices.

When India exploded its "peaceful" nuclear device in 1974, the global refocus on nuclear non-proliferation which followed forced the government of Zulfikar Ali Bhutto to enlist the assistance of Khan in avoiding the tightening of global controls over the dissemination of the nuclear technology Pakistan needed, as its nuclear weapons program developed. Competition between Pakistan and India, present since the birth of the state of Pakistan in 1947, now gained a nuclear dimension.

---

6 Unless otherwise noted, information in this section is derived from:

Albright, D. and Hinderstein, C., The A.Q. Khan illicit nuclear trade network and implications for nonproliferation efforts, *Strategic Insights*, Vol. 1, Issue 6, (July 2006). Available online at: http://www.ccc.nps.navy.mil/Si/2006/Jul/albrightJul06.asp. Accessed May 9, 2009, and: *International Institute for Strategic Studies*, Nuclear black markets: Pakistan, A.Q. Khan and the rise of proliferation networks, a net assessment: Chapter 3: *A.Q. Khan and onward proliferation from Pakistan*, (May 2, 2007). Available online at: http://www.iiss.org/publications/strategic-dossiers/nbm/. Accessed May 9, 2009.

A.Q. Khan's early activities in developing his country's uranium enrichment technologies may well have remained narrowly focused on that national goal as he and his growing organization became increasingly enmeshed and skilled in the means and methods of illicit nuclear technology procurement. These included the use of Pakistan's diplomatic corps, the practice of paying over market price to obtain needed items, intensive search techniques seeking multiple supplier corporations, the use of "front" corporate identities, changing end-user documentary information, altering specifications to avoid scrutiny by nuclear regulators, and shipping items through free ports, such as Dubai, with few controls over the movement of goods.[7]

However, the degree of autonomy which Khan and his organization enjoyed, the need for high levels of secrecy as Western and other intelligence agencies sought information on Pakistan's nuclear aspirations, and rivalries between Khan and other Pakistan nuclear organizations all contributed to his determination to engage in further proliferation beyond his own country. Simply put, A.Q. Khan's organization was open for nuclear business, world wide.

In reality, this meant doing business with Iran, North Korea and Libya for the full period from the mid-1980s to the time of the suppression of Khan's network in late 2003. Perhaps the most significant element of Khan's proliferation activities lay in his provision to customer states of "shopping lists" of suppliers of many of the components and technologies necessary for building completed nuclear weapons, rather than undertaking his own direct supply from surplus Pakistani stocks. This is known to have occurred in regard to Iran's receipt not only of complete gas centrifuges but also of components and technical design information. Again, although Khan's organization is known to have supplied North Korea with over twenty centrifuges (far too few to produce sufficient enriched uranium for a nuclear explosive device) he also provided the technical data and drawings which were vital in the indigenous construction of many more.

In these dealings, it remains an open question as to whether the Pakistani government was complicit, or at least acquiesced, in their completion. It appears, from a distance and with the benefit if hindsight, that there must be at least a reasonable possibility that Islamabad and Pyongyang cooperated to allow Khan's activities in respect of North Korea to proceed.

In the case of Libya, which hitherto had developed no significant nuclear capacities, the Khan network appears to have acted purely as an organization seeking to profit from straightforward business dealings. Libya was seeking a fully supplied and complete gas centrifuge uranium enrichment plant, to be fully supported technically by the network. The Khan network sought to comply with

---

7 John Chipman, Director General and Chief Executive, International Institute for Strategic Studies, Press Statement, May 2, 2007, ISS Strategic Dossier, *Nuclear Black Markets: Pakistan, A.Q. Khan and the rise of proliferation networks – A Net Assessment.* Available online at: http://www.iiss.org/publications/strategic-dossiers/nbm/nuclear-black-market-dossier-press-statement/. Accessed May 9, 2009.

these requirements by bringing together the plant's components from commercial suppliers in Malaysia, Turkey, South Africa and in Europe, without the obvious involvement of the Pakistan government. The components were to be transshipped through Dubai. Although this operation unraveled in October, 2003 with the seizure of centrifuge parts on board the German-registered ship *BBC China*, en route for Tripoli via Malaysia and Dubai, the Khan network's complexity and global capabilities were clear. Most concerning of all was the fact that, when the network was closed down, Libya confirmed that it had received technically high-level documents and drawings for the construction of nuclear explosives. These covered nuclear weapon theory, the design details of individual components, and almost all the drawings necessary for an implosion-type bomb design. Allegedly of early Chinese origin, the design has been described in 2006 as representing an entirely feasible, well engineered bomb design.[8]

The A.Q. Khan network contained within it the potential to swiftly initiate a period of unstoppable nuclear proliferation in which a range of unstable and failing states, as well as terrorist organizations with nuclear ambitions, could find themselves separated from operational nuclear devices only by the money necessary to access them. That this nightmare situation has been partially avoided is in large measure due to the work of several Western intelligence agencies, although there is still no certainty, for example, about whether the Khan network supplied other customer states. In this respect, there remains speculation that it undertook preliminary negotiations for the supply of gas centrifuges with agents of Saddam Hussein's Iraq from the Mukhabarat intelligence service during late 1990.[9]

It was only with the advent of the era of the Khan nuclear smuggling network that credible threats emanating from the misuse of nuclear fuels left the exclusive domain of state control and entered the realm of private and commercial illegal nuclear activities. Until then, incidents of non-state nuclear malfeasance had been both ad hoc and transitory, rather than complex, highly organized, or on a scale relevant to bilateral negotiations with the governments of nation states.

The success which Khan and his associates enjoyed was the result of his ability to act both as a representative of the government of a sovereign state *and* as a businessman, emphasizing one or the other as circumstances dictated. In doing so, he was energized primarily by the profit motive, which he pursued within a political environment which allowed him the opportunity to leverage his privileged position with the government of Pakistan for private gain.

State-based Diversion of Fissile Materials

Over the sixty-five years of the nuclear era, popular and specialist visions of the dual and simultaneous possibilities of both unlimited energy and devastating

8 Supra. At Note 6 (Albright and Hinderstein).

9 Ibid.

nuclear war, present from the beginning, have grown and receded by degrees as the development of new technologies and the recurrent instability of international affairs have in turn commanded the world's attention. Would the promise of seemingly limitless nuclear energy, sufficient to fuel the human future, ever be fulfilled as its advocates asserted, especially in the face of alternative and new sources of renewable energy? Conversely, would the indescribable horror of a nuclear exchange be the inevitable result of global confrontation between two or more large powers, armed (as they were and continue to be) with multiple thousands of operational, nuclear-tipped intercontinental ballistic missiles capable of destroying life on Earth?

The world's first coordinated response, led by the United States and the Soviet Union in negotiations at the Eighteen Nation Disarmament Committee in Geneva during 1967–68, was the multilateral Nuclear Non-proliferation Treaty, which came into force in 1970. This global bargain between those states with, and without, nuclear weapons is the central pivot around which the nuclear non-proliferation regime turns. Acknowledging the dual threat and promise of nuclear energy, the NPT legitimated an historic compromise which allowed nuclear energy for the benefit of human societies to flourish while attempting to ensure that it could not be diverted for use in weapons of war.

Those states without nuclear weapons – which amounts to all states except the five Nuclear Weapons States and the four de facto nuclear powers Israel, North Korea, India and Pakistan – progressively signed on to a bargain which held out hope that the nuclear weapons genie could be put back in the bottle through the cooperative efforts and enlightened self-interest of all states.[10] At the same time, the promise of the peaceful use of nuclear energy for the future would, it now appeared, be available on a non-discriminatory basis to all states including, of course, the established nuclear weapon states themselves.

It is an axiom of international affairs that those agreements and treaties in the field of military conflict between states which do not include mechanisms to verify the statements and actions of states in fulfillment of their obligations are at best unstable, at worst dangerous. In the case of the NPT, the cornerstone of nuclear non-proliferation since its inception, this truism is an indispensable *leitmotif*. In this instance, it must have appeared improbable to its American and Russian negotiators in Geneva that no state without nuclear weapons would ever seek to acquire them through diversion of legitimately held civilian fissile material to military use in a nuclear weapon program.

In any event, this was not a risk the world could take, meaning that a regime of inspection and verification of state adherence to obligations undertaken under the

10 North Korea acceded to the NPT on August 4, 1985 after Russian pressure to do so, but announced its withdrawal from the treaty in April 2003 following the failure of the 1994 Agreed Framework between the DPRK and the United States. North Korea exploded its first nuclear device on October 9, 2006. Israel, India and Pakistan have not signed or ratified the NPT since its entry into force in 1970.

terms of the NPT would be a vital part of the NPT nuclear contract, and its shape appears in Article III. In essence, this seminally important element of the NPT bargain requires all states party to the treaty to conclude a Safeguard Agreement with the International Atomic Energy Agency verifying through the inspection activities of the agency that no diversion of fissile material from peaceful use to nuclear weapons or devices has taken place. In addition, each state party agrees not to transfer fissile material, or the capacity to produce it, to other non-nuclear weapons states unless those states are themselves under the scrutiny of the IAEA's inspection and verification regime.

Thus, the NPT distills in this article both the central dilemma of nuclear non-proliferation and a potent response: the bargain between the nuclear weapons "haves" and "have-nots" is to be upheld by the simple aphorism "trust but verify".

The Nuclear Fuel Cycle

At the time of writing there are 436 nuclear power reactors in operation in thirty-one countries, with a total net installed capacity of 370 gigawatts electricity (GW (e)), while forty-five new reactors are under construction. Nuclear power provides around 16 per cent of electrical energy generated worldwide, although some 80 per cent of nuclear generating capacity is concentrated in only eight countries: the United States, France, Japan, Germany, Russia, South Korea, Ukraine, and Canada. As world wide demand for low carbon emission energy sources continues to expand, installed nuclear generating capacity is expected to grow, although estimates of installed capacity by 2030 vary across a range from 400 to 600 GW (e). If nuclear power generation continues to develop in response to further demand beyond 2030, a Massachusetts Institute of Technology interdisciplinary study has projected installed capacity of around 1,500 GW (e) in fifty-eight countries by 2050.[11]

It appears defensibly predictable, then, that nuclear powered electricity generation will continue as a significant and growing contributor to world energy resources in the decades ahead. Whether or not that actually occurs, in the face of high capital costs, uncertain energy demand growth, scarcity of capital in developing countries, and widespread social resistance, is far less certain. Whatever the final outcome, though, there is no doubt that the presence of nuclear power generation industries around the world carries with it the threat that, in one way or another, the fuels they use may be diverted away from legitimate use, and towards clandestine nuclear weapons development programs.

---

11 International Panel on Fissile Materials, Global Fissile Material Report 2007, *Second report of the International Panel on Fissile Materials developing the technical basis for policy initiatives to secure and irreversibly reduce stocks of nuclear weapons and fissile materials*. Available online: http://www.fissilematerials.org. Accessed May 14, 2008. At pp. 82–85.

To understand the scope of such threats, it is necessary to acknowledge the two primary points of vulnerability within the nuclear fuel cycle. At the front end of the cycle, as enriched uranium fuel is fabricated, the primary threat lies in the development of nationally controlled uranium enrichment plants, the possession of which confers on states immunity from international withdrawal of supply. (As always, fissile material lies at the heart of all nuclear threat scenarios.)

At the back end of the fuel cycle, after the energy potential of the uranium fuel has been functionally exhausted, the danger lies in the ultimate disposition of the plutonium present in spent reactor fuel. Approximately 10,000 mt of spent reactor fuel is discharged each year around the world, incorporating some 75 mt of plutonium, in the form of its isotopes Pu239 and Pu240. If separated from spent fuel by recycling, the isotope plutonium-239 can be readily mixed with low enriched uranium and used as mixed oxide ("MOX") fuel in the many Light Water Reactors around the world, or as the fissile material in nuclear weapons.

The Enriched Uranium (HEU) threat

Taking the front end threat first, uranium enrichment capacity is currently concentrated in Russia, the United States, Western Europe and (recently) in Japan. Naturally, a rapid rise in world demand for nuclear power generation will result in a corresponding rise in demand for uranium enrichment capacity. The International Panel on Fissile Materials, an independent group of arms-control and nonproliferation experts from sixteen countries, calculated in 2007 that if the MIT estimate noted above of 1,500 GW (e) nuclear generating capacity by 2050 is realized, it would result in a five-fold increase in demand for enrichment services.

It would be dangerously unrealistic not to expect that a significant proportion of this growing demand would be met by the construction of new national enrichment capacity, especially by countries new to nuclear power, and perhaps with ambitions towards a nuclear weapons program in the future. Although countries presently enriching uranium could satisfy such a rise in demand through their own resources, concerns over future energy security alone may be sufficient to convince a number of new nuclear players that independence in nuclear fuel supply is essential for their future economic success. Around 20 countries are forecast by MIT to develop generating capacity of at least 10 GW (e) by 2050, including Iran, Pakistan and Indonesia, while some 30 countries not currently generating nuclear power will require enrichment services by that date.[12]

The pivotal question is how countries in the latter category (and in the final analysis *all* nuclear power generating countries) should obtain uranium enrichment services. In this book, Chapter 8: "Old and New Strategies for Non-proliferation Progress" will address the range of options in greater detail, but it is useful here to consider the general scope of possible responses. The immediate aim would be

12 Ibid. At pp. 83–91.

to deny to any country seeking to build enrichment facilities the technological and material means to do so while providing an acceptable alternative to those countries whose need for enriched uranium is linked solely to its search for future energy security. In building the tools to accomplish such a feat, the need to construct a multilateral response would present an opportunity to put in place a globally coordinated system of uranium enrichment. Furthermore, such a system would be managed by, and operate under the aegis of, the International Atomic Energy Agency, whose expertise, international legitimacy and impartiality is essentially unchallenged around the world.

In pursuit of such a goal, the retiring Director General of the IAEA, Dr Mohamed ElBaradei, has proposed a stepped approach in which several important and difficult steps would be necessary. *First*, there would be a need for a system assuring and guaranteeing the supply of fuel for all reactors, and of the reactors themselves where necessary. *Second*, all new enrichment (and reprocessing) facilities would come under immediate and complete international control. *Third*, all existing facilities would be converted to international control. *Finally*, a prerequisite would be the conclusion of a long proposed but thus far elusive multilateral Fissile Material Cut-Off Treaty (FMCT) prohibiting the production of fissile material for nuclear weapons.

In setting out the process by which the world can avoid the threats posed by uncontrolled uranium enrichment, Dr ElBaradei offered his design for the key elements of an enriched uranium supply mechanism. In four parts, he proposed that:

- The supply mechanism must operate in an unambiguously multilateral environment which is not dominated by the current supplier states or major powers. Instead, and in order to gain the confidence of countries considering a nuclear power program, it must consist of an equal partnership of supplier and consumer states.
- A keystone would be the equal status of all participant states in their ability to access fuel enrichment services.
- The release of fissile material to a consumer state must be based on non-political criteria, which would be established in advance and applied in an objective and consistent manner.
- The context in which such a scheme will succeed will be one in which it forms part of an overarching multilateral framework. This would be one in which nuclear disarmament is acknowledged by all nuclear weapon states, but especially the United States and Russia, as an integral component of global nuclear non-proliferation efforts.[13]

13 International Atomic Energy Agency, Statements of the Director General, April 17, 2008, *Nuclear energy: the need for a new framework*. Available online at: http://www.iaea.org/NewsCenter/Statements/2008/ebsp2008n004.html. Accessed May 17, 2009.

The increasingly broad interest in "fuel bank" proposals of this kind has resulted in no fewer than twelve such proposals to the IAEA since 2003, seeking to institute systems or protocols with elements of ElBaradei's vision, and these will be examined in later chapters. It is significant that many of them envisage a central role for the IAEA.

The spent fuel reprocessing threat

As noted earlier, the fuel rods discharged from nuclear reactors, following the exhaustion of most of their fissile uranium, contain a wide range of radioactive fission products, a small amount (usually less than one percent) of unfissioned uranium-235, nearly all of the non-fissioning uranium-238 present in the fuel rods when new, and up to one percent of fissile plutonium. In an open-ended nuclear fuel cycle, no reprocessing of spent fuel is undertaken, and the spent fuel rods are simply cooled for several years in water ponds before being transferred to permanent dry-cask storage facilities, both usually located on the reactor sites themselves.

While this is a relatively low-cost fuel strategy option, much potential energy remaining in the spent fuel is thereby lost, while the capacity which reprocessing has for reducing the quantity, longevity and radiation levels of spent fuel is ignored. The security implications inherent in storing spent fuel containing unseparated plutonium has thus far been considered manageable, largely due to the presence in spent reactor fuel of highly radioactive and dangerous radionuclide fission products.

However, reprocessing of spent fuel to recover its fissile plutonium and uranium can extract some 25 per cent additional energy from the original uranium, as well as reducing by up to 30 per cent the quantity of natural uranium feedstock which would otherwise be required for newly fabricated fuel. The economic benefits to be gained from reprocessing are, however, complex, being dependent, for example, on the world price of natural uranium and the low relative cost of nuclear fuel as a component of total generating costs. Even so, the fact that reprocessing adds cost to completed nuclear fuel, when compared to fuel processed from natural uranium ores, is not disputed, and is difficult to justify on strictly economic grounds alone. Incentives for undertaking reprocessing, however, include a reduction in the volume of high-level nuclear wastes requiring long-term storage by around 80 per cent, as well as the possibility of using the separated plutonium in future fast breeder or burner reactors. Simply put, there are significant advantages and disadvantages in reprocessing nuclear fuels.

However, there is no doubt that the separation of fissile plutonium from spent reactor fuel rods for use in new nuclear fuel presents a clear proliferation threat to the extent that the plutonium thereby produced becomes, at least notionally, accessible to nuclear terrorists, or to proliferating states. The threat is ameliorated through the incorporation of plutonium in mixed oxide (MOX) fuel, although again the picture is muddied by the fact that the cost of storing the high level waste and MOX spent fuel is roughly equal to the cost of merely storing spent fuel directly upon its discharge from the reactor.

Nevertheless, reprocessing of spent reactor fuel is a long-established strategy for dealing with, and gaining advantage from, waste nuclear material, although the disadvantages inherent in reprocessing mentioned above have led a number of countries to cease shipping spent fuel to established reprocessing facilities, such as Rokkasho in Japan and La Hague in France. Some ten or more countries which previously sent spent fuel to France, Russia or the UK for reprocessing have in recent years ceased doing so, impelled by contemplating the difficulties involved in finally having to repatriate and store the separated plutonium and radioactive waste products on their own territories, as they have agreed to do. By contrast, the United States, Japan and France plan to initiate or continue reprocessing significant percentages of their own spent fuel, although the United Kingdom plans to discontinue reprocessing from 2012.

Wherever spent reactor fuel is reprocessed to extract its plutonium load (amounting to some 2,000 mt per annum) security concerns emerge, although national strategies for dealing with spent fuel through reprocessing vary. France extracts plutonium once, uses it to create MOX fuel, and then stores the depleted MOX for possible use in fast breeder reactors of the future. The United Kingdom currently reprocesses its spent fuel, storing the plutonium thus produced, but intends to cease reprocessing completely and permanently by 2012. The United States abandoned reprocessing in the late 1970s, but now appears to be on a path towards large-scale reprocessing with the 2009 abandonment by the new Obama Administration of the proposed long-term repository at Yucca Mountain, Nevada. Each strategy holds implications for the security of fissile material capable of use by nuclear terrorists, who will not hesitate to take full advantage of what at the moment is a fragmentary and uncoordinated global fuel security situation.

Fuel Security Policy and Practice

In the present era of global fiscal stringency and uncertain economic futures, the capacity of the governments of all countries to respond to growing calls on the public purse is necessarily constrained. When combined with unrelenting demands across the international community for enhanced commitments to international security cooperation, the many claims on public sector finances – all competing for favoritism and special treatment – limit policy spending options in terms of priorities, timing, efficiency of delivery and effectiveness. This state of affairs is an unavoidable and necessary condition of national governance at any time, but even more so during difficult economic times.

Even so, the place of national security policy within a hierarchy of unlimited national and international needs and demands, when arrayed against limited public resources, should be, and almost always is, at or near the top of the list. No national government worthy of the name can afford to canvass, in the public domain, any alternative and hope to survive at the ballot box. It is the first and primary duty of all governments to provide for the defense of the nation as the *sine qua non* of all other, and secondary, considerations.

In the case of the United States, the Homeland Security Act of 2002 defines "Homeland Security" as " … the deterrence, prevention, and preemption of, and defense against, aggression targeted at US territory, sovereignty, population, and infrastructure as well as the management of the consequences of such aggression and other domestic emergencies"[14] Surely no demand on the sum of national resources could reasonably be considered more important or grave, especially in the present international political climate of serious and growing concern about the threat of international nuclear proliferation and terrorism.

It follows from this finding that those elements of public policy which are most capable of succeeding in securing a country's territory, sovereignty, people and physical assets against clear and credible threats against them must be the primary focus of all and any national security policy, and of its effective practice. National security policy, of course, covers a vast continent of ideals, histories, personalities, national traits and talents, geographies, politics, economic, technical and military capacities and potentials, and international alliance relationships, to name but a few of its many parts. Each of these broad categories, and many more, also form integral elements within the narrower field of nuclear security policy, as well as the even more focused subject of this discussion: nuclear fuel security policy and practice.

The primary challenge for planners of future national and global nuclear fuel security protocols, processes and regimes (the two fields being closely linked) must be how best to attain a closely coordinated global fuel security strategy which is capable of attracting the ultimate support of all states, and initially that of all nuclear and nuclear-interested states. The final goal must be to reach a position in which all states party to the Nuclear Non-proliferation Treaty, as well as those not currently member states (Israel, India, Pakistan and, de facto, North Korea), enter into a multilateral treaty providing for the international control and final disposition of all nuclear fuels, both extant and prospective.

Modeled on the principles and practical effects of the NPT, and intended as a coordinated complement to it, such a treaty must incorporate comprehensive obligations, imposed equally on all states, to submit to a stringent inspection and verification regime to be undertaken by the Inspectorate of the International Atomic Energy Agency in respect of the disposition of all nuclear fuels used to generate electrical energy, or for any other legitimate and peaceful purpose.

It would be naïve and unhelpful to expect that this ideal state could be reached in the short term, or that its longer term institution would not be attended by the inevitable diplomatic twists and turns accompanying such a universal goal. Even so, the history of the negotiation of the NPT during the latter years of the 1960s

---

14 For an expanded definition of Homeland Security, refer to Chapter 2, at "Homeland Security: A Vision" at pp. 31–2.

provides a precedent for success when the need is sufficiently pressing. During that era, the atmospheric testing of nuclear and thermo-nuclear devices between 1945 and 1963 resulted in a rapid rise in the level of background atmospheric radiation levels which reinforced the dire consequences of nuclear proliferation in the minds of the American and Soviet governments of the day.

President John F. Kennedy spoke to the American people in 1963 on the Nuclear Test Ban Treaty in 1963 in these terms:

> I ask you to stop and think for a moment what it would mean to have nuclear weapons in so many hands, in the hands of countries large and small, stable and unstable, responsible and irresponsible, scattered throughout the world. There would be no rest for anyone then, no stability, no real security, and no chance of effective disarmament. There would only be the increased chance of accidental war, and an increased necessity for the great powers to involve themselves in what otherwise would be local conflicts.[15]

President Kennedy's vision is particularly apposite since the world seems now to be approaching a nuclear proliferation tipping point which may only prove manageable, if indeed it is manageable at all, through the international control of nuclear fuels. The ability of the international community to control the disposition of nuclear fuels should not be doubted, but it can only be accomplished with the strongest possible levels of coordination and cooperation of the what, how, where and when of nuclear fuel security.

It is no coincidence that the single international organization capable of performing this task has for a number of years advocated this very goal. The International Atomic Energy Agency, headquartered in Vienna, Austria, has the necessary technical capacity, international legitimacy and universal confidence of the international community in its impartiality, if not the current resources to successfully undertake this immensely important element of its nuclear non-proliferation work. Later chapters will develop a proposal for the enlistment of the IAEA as leader, coordinator and executor of global nuclear fuel security strategy. It remains here to undertake a preliminary assessment seeking to prioritize responses to the suite of fuel-related proliferation threats, large and small, which combine to focus the attention of national and international security planners around the world.

---

15 Carnegie Endowment for International Peace, *JFK on nuclear weapons and non-proliferation*, November 17, 2003. Available online at: http://www.carnegieendowment.org/publications/index.cfm?fa=viewandid=14652andproj=znpp. Accessed May 25, 2009.

Responding to Fuel Threats: A Simple Triage Approach

In the final part of this chapter, the policy challenges and options open to the United States (and indeed all nuclear-interested states) in confronting nuclear fuel security threats will be introduced. The range of available and alternative US policy options will be used to develop a tripartite multilateral approach to prioritizing and coordinating policy which seeks to react effectively and efficiently to multiple fuel threat scenarios. Suggested prioritizing criteria to be applied are:

- the nature of a specific threat
- its temporal and effective proximity
- the severity and durability of its anticipated effects
- the full cost of an effective and efficient response
- the length of a reasonable time frame for response implementation.

In order to indicate the way in which this triage approach is applied for the full suite of nuclear fuel threats (which will be further elaborated in later chapters) the disposal of excess US and Russian plutonium, from 2000, through formal agreements between the two states signed in 2000 and 2007, will be used as an exemplar. The delayed implementation of this program in the face of diplomatic and other hurdles is emblematic of the difficulties to be anticipated in undertaking and completing such a sensitive and complex process.[16]

In purely medical terms, simple combat triage is a method of allocating finite resources in the most efficient and effective manner possible. By dividing wounded soldiers into three categories, priorities for the removal of victims of injury from the battlefield, and for their subsequent care, are determined by the category assigned to them. These categories are:

- those for whom immediate care could enhance their chances of survival – "Emergency"
- those who are likely to live, regardless of the care they receive – "Urgent"
- those who are likely to die, regardless of the care they receive – "Non-Urgent"

The complex medical situation confronting emergency physicians dealing with the early arrival of wounded soldiers from a battlefield is mirrored in the complex and dynamic global nuclear non-proliferation and security picture. Whether the focus is on civil or military nuclear affairs, the nuclear fuel cycle or international strategic

16 Unless otherwise indicated, data used in this section is from The Nuclear Threat Initiative, *NTI: Issue Brief, Plutonium Disposition*. Available online at: http://www.nti.org/e_research/e3_11a.html. Accessed May 27, 2009, and US Department of Energy, *News Archive, October– December 2007*. Available online at: http://www.energy.gov/news/5742.htm. Accessed May 27, 2009.

security, rapidly developing threats and innovative responses sit side-by-side with decades-old nuclear questions and their equally indeterminate solutions.

Thus, the emergence of clear evidence of Libya's nascent nuclear weapons program in 2003, following the interception of a shipment of Malaysian-built gas centrifuges, resulted in Libyan President Muammar Gaddafi's rapid closure of his country's uranium enrichment program. Conversely, the continuing interest demonstrated by India in the use of thorium as a substitute for uranium as a nuclear fuel could pose new security questions in the medium term, should India succeed in creating a viable thorium fuel cycle. Thorium is some five times more abundant than uranium in the Earth's crust, and many countries have ample supplies. Furthermore, thorium used in a modified light water reactor is capable of producing the fissile isotope uranium 233, which can be readily used in simple, gun-type nuclear weapons.

The international community is faced with a multitude of simultaneous nuclear proliferation issues and threats. Some can be measured in a scale of one or two years, others in decades; some have simple and clear solutions, some are as intractable today as they were twenty or more years ago. All form part of the nuclear non-proliferation question and its comprehensive solution.

The world has no choice but to address and deal with each of these disparate problems in the most efficient and effective manner possible, if there can be any hope of successfully confronting those who seek to use nuclear technology in malign ways. In a general sense, nuclear non-proliferation policy and its implementation must increasingly be coordinated in a coherent way which focuses primarily on attaining, in a sequential manner, identifiable, measurable and feasible goals.

When targeted specifically at nuclear fuel security policy, however, the benefits available from the triage approach become more readily and practically apparent. Examples among a large suite of extremely urgent measures are manifold for analogues of each of the three simple triage categories noted above. The "Emergency" category of nuclear fuel security issues – those in which immediate or continued attention should enhance an outcome – include the blending down by the United States and Russia of their excess weapons-grade highly enriched uranium (HEU) to low-enriched uranium suitable for use in light water production reactors.

Can such an immediate priority as this be assigned to this particular issue? The five point list of prioritizing criteria noted above suggest that it can:

- "*the nature of a specific threat*": the presence of excess weapons grade HEU in Russia represents the clear threat of theft by terrorists or unauthorized removal by Russian personnel, given the continuing deficiencies still present in Russian physical security arrangements.
- "*its temporal and effective proximity*": the HEU in question, in both the US and Russia can, with goodwill on both sides, be removed from harm's way both rapidly and completely, and subsequently be removed permanently from the risk of diversion.

- "*the severity and durability of its effects*": the catastrophic effects of the successful use of this nuclear material, in a crude uranium bomb by a terrorist organization, is as obvious as its immediate and longer term effects, whether individual, political, social, economic or otherwise.
- "*the full cost of an effective and efficient response*": the quantifiable cost measured in resources expended is relatively negligible; the opportunity cost of failing to carry out the program could be incalculably immense.
- "*the length of a reasonable time frame for response implementation*": although the removal and processing of HEU presents some significant issues in handling and transportation, these are well understood in both countries, and can be accomplished in a relatively short time – dependent, of course, on the quantities involved.

It appears open to claim that an exercise of this kind is adaptable to most, if not all, of the many definable security issues inherent in the manufacture, use, re-use and disposal of nuclear fuels. Although this discussion is merely an introduction to the analytical basis of more complex policy development, it is nonetheless a crucially important one.

Nuclear fuel policy issues may be similarly placed in the "urgent" category using these prioritizing criteria. They may be characterized as those requiring attention and resolution in order to allow for steady and controlled progress in countering the further diffusion of uncontrolled nuclear technologies and materials. Examples include new, more proliferation-resistant Generation IV reactor technologies (to be discussed in later chapters) which deny proliferators access to separated plutonium after reprocessing, as well as the international and informal Proliferation Security Initiative to halt nuclear smuggling.

Finally, in the "Non-urgent" category lie globally significant but distant re-orderings of the international system in ways which exert fundamental and novel changes in the production and consumption of energy, and specifically of nuclear energy and its use in weapons of war. Examples might include complete nuclear weapons disarmament, operational nuclear fusion technologies and a renovated and rejuvenated United Nations Organization.

Having discussed and characterized the nature of threats to nuclear fuels, and a range of necessary responses, this analysis now turns to the historical landscape of success and failure evident in the sixty-year-old field of securing the nuclear fuel cycle and nuclear weapons fissile materials. In particular, it examines the record of deficiencies apparent in safeguarding the uranium and plutonium stages of the cycle, and the consequences that have followed. The uneven and often inadequate global response to the clandestine use of nuclear fissile materials is emphasized in terms of opportunities unrecognized and lost, in the hope that this exercise will provide useful strategies for avoiding the pitfalls of the past while reinforcing success.

Chapter 5
The Historical Record: Failure and Success

Introduction

When the first atomic bombs descended by parachute over the Japanese cities of Hiroshima and Nagasaki, Japan early in August 1945, the properties of both types of fissile material within them were well understood in the United States and beyond. The 130 lbs of uranium in the Hiroshima weapon, a gun-type device known as "Little Boy," had been highly enriched in U-235, its main fissile isotope, at enormous financial cost and expenditure of research effort by the United States (with the help of the United Kingdom and Canada) as part of the Manhattan Project to develop nuclear weapons and end the Second World War. "Little Boy" produced a blast yield equivalent to around 13 kilotons of TNT, while its counterpart at Nagasaki, the "Fat Man" implosion device, used some 14 lbs of plutonium-239 to produce a blast yield through nuclear fission equivalent to approximately 21 kilotons of TNT.

At Hiroshima, at least 140,000 people had perished by the end of 1945 as a result of the explosion, with many more subsequent deaths in the city attributed to the longer term effects of exposure to ionizing radiation. In Nagasaki, the number of immediate and short-term fatalities was estimated at 80,000. The two explosions were extremely small by modern standards, being one order of magnitude smaller than later fission weapons, and at least three orders of magnitude smaller than the later US thermonuclear ("H-Bomb") weapons of up to 25 megatons (million tons) of TNT equivalent explosive power.[1]

Thus, at the very beginning of the modern nuclear era, the two primary fissioning nuclear materials still in use today, uranium-235 and plutonium-239, had been deployed in nuclear explosive devices whose effectiveness had been demonstrated in the most dramatic and appalling way. These two crude nuclear weapons of sixty-five years ago, detonated over Japanese population centers, set a global course for nuclear energy evolution which simultaneously promised

1 Throughout this book, the term "nuclear weapon" is used to describe a simple gun or implosion fission device, as described earlier and used over Japan in 1945. Devices described as "thermonuclear weapon," which normally use isotopes of hydrogen (deuterium or tritium) to boost their fission energy through the nuclear fusion of these and other light isotopes, are usually far more powerful than simple fission weapons, but also far more difficult to construct (and impossibly so for nuclear terrorist organizations). This does not, of course, preclude their theft, or other forms of their illicit use.

humanity both unequaled economic and social benefits and the unimaginable tragedy of generalized nuclear war.

The immediate intellectual challenge for a world exhausted by global conflict was embedded in the need felt by United States administrations of the period to permanently quarantine, control and safeguard the technology of nuclear armaments, denying its potential to all comers while preserving its ultimate guarantee of America's global strategic ascendancy. At the same time, the nuclear promise of virtually unlimited energy from the splitting of the atom provided the means by which an emergent and aspirant middle class, in the US, the United Kingdom and later in Western Europe, could build their hopes of lasting economic success from transient military conquest. Although the story of America's immense economic growth and social transformation during the nineteen fifties and beyond, built largely on accelerating domestic consumption and abundant energy, lies beyond the scope of this discussion, its lessons were not lost on her allies and adversaries. The conundrum confronting US nuclear policy planners thus became how to safely allow the rest of the world to share in the undoubted benefits of America's nuclear power technologies, while denying any possibility of the diffusion of nuclear weapons or materials, since the fundamental technologies and materials necessary for each were essentially the same. In all of this, and from the very beginning, the availability of fissioning nuclear material for weapons of war or electrical power was the obvious lever of influence, and remains so today.

This chapter explores the landscape of failure and success in the decades old field of securing the nuclear fuel cycle against misuse and exploitation, leading to nuclear proliferation and the possibility of nuclear terrorism. Taking note of views expressed throughout the nuclear era by, and in respect of, nuclear-interested agents such as the International Atomic Energy Agency, the governments and agencies of nuclear and nuclear interested states, non-governmental non-proliferation advocacy organizations, other specialist think tanks and the broader international community, it examines the record of deficiencies apparent over the years in safeguarding the front and back ends of the nuclear fuel cycle, and the consequences that have followed. Its purpose is to highlight a range of successful and failed global and regional nuclear fuel policy responses to security concerns, and to assess the effectiveness of their global interpretation and application. In doing so, it is intended to provide a concise means to assist strategic nuclear policy planners and their advisors in avoiding the broad strategic nuclear planning pitfalls of the past, while reinforcing success.

Examples used to illustrate this process include North Korea's initiation of a weapons development program from the mid-1950s, and India's clandestine development of nuclear weapons from the early 1960s. A small case study is also offered in respect of Pakistan's protracted move to acquire nuclear weapons technologies and materials. This third example is highlighted in light of Pakistan's long-held animus towards its nuclear-armed neighbor India, together with its pivotal role in regional security, its unstable political past and present, and the presence of terrorist groups within its borders.

The uneven and often inadequate global response to fuel security leading to nuclear weapons proliferation is emphasized in terms of opportunities lost amid the consequences of the continuing proliferation of nuclear materials and technologies.

An Historical Overview

It is helpful at the outset to re-emphasize the fact that uranium enrichment and spent nuclear reactor fuel reprocessing present the two most serious of all challenges to the nuclear non-proliferation regime.

Both processes, legitimately part of the means by which nuclear electrical energy is generated, also directly produce fissioning weapons-grade nuclear material in the form of highly enriched uranium and separated plutonium. The pathways through which the sixty year history of nuclear fuel security evolved were built on these simple realities, the consequences of which have constantly occupied the minds of United Nations nuclear bureaucrats, their counterparts within national governments, the leaders of industrial corporations making their way within private nuclear energy sectors, and the many components of civil society at the individual, group, national and global levels.

It is perhaps not surprising that a novel, indeed a universal approach was regarded, in the aftermath of the Second World War, as essential when confronting the need to establish global rules governing the use of nuclear energy. The first expression of this approach was the United Nations Atomic Energy Commission (UNAEC), founded on January 24, 1946 by the very first Resolution of the UN General Assembly, and charged with reporting to the Assembly on the "international control of atomic energy". The UNAEC had been called into existence by the United States, the United Kingdom and the Soviet Union shortly after the end of the Second World War in an atmosphere of urgency, as the three primary victor states sought international solutions to the need to control, limit or indeed destroy all extant nuclear weapons, while working towards progress in the peaceful application of nuclear physics. Its membership initially consisted of the five original permanent members of the Security Council (the United States, United Kingdom, the Soviet Union, France, and the Republic of China) with Canada as an additional member state. Within three years, the General Assembly had expressed its deep concern over the impasse reflected in the Commission's three initial reports on the international control of nuclear energy. By 1949 the Commission, having failed to reach any substantive agreement among its member states, had adjourned indefinitely.

The only successful feature of this first, failed attempt at global nuclear agenda and rule setting was the appearance of an American plan, presented to the UNAEC in 1946 by the US delegate on the Commission, Bernard Baruch, for the introduction of a different kind of international bureaucracy: the Atomic Development Authority. The ADA would, claimed Baruch, oversee the development and use of nuclear energy, manage all nuclear weapon production facilities, and have the right to inspect all peaceful use nuclear facilities. In addition, "illegal" weapons were to be prohibited, while any facility under the aegis of the ADA could be seized, and those interfering with its inspections regime punished. The wider United Nations context would be that the Security Council could impose sanctions on any state which refused to abide by the terms of the plan. However, and most significantly of all, the member states of the Council would lose their power of veto over any punishment to be meted out to state violators. Finally, on full implementation of the plan the United States would begin the process of destroying all of its nuclear weapons holdings.[2]

With the hindsight available more than six decades later, it seems clear that the Soviet rejection of such a plan was inevitable in the shadows of the early stages of the Cold War. To allow America to retain its (then) nuclear monopoly, with no guarantee of its willingness to engage in full nuclear disarmament and, further, to relinquish UN Security Council veto power, were two bridges too far in Josef Stalin's Moscow. Perhaps the greatest surprise, at least in today's world, is that the Truman Administration saw any hope that the Baruch Plan's rigid and prescriptive approach to nuclear *rapprochement* and disarmament, dictated to the world and backed by America's global military supremacy, would be accepted without demur by any of its major allies or potential foes.

By 1953, a more optimistic and conciliatory diplomatic atmosphere prevailed as President Eisenhower gave his famous "Atoms for Peace" speech to the United Nations General Assembly on December 8, 1953. Certainly, though, in the intervening seven years since the demise of the Baruch Plan, the urgency of the need to curb the production and proliferation of nuclear weapons had grown rapidly. To this end, Eisenhower's proposals to establish an "international atomic energy agency" to act as a repository for the "normal uranium and fissionable materials" of all states reflected a move in US Government circles towards cooperative non-proliferation effort rather than any Washington *fiat*.

Furthermore, fissionable nuclear materials would be made available for allocation to the "peaceful pursuits of mankind," marking the first appearance of the notion of an international fuel bank, to be operated under international guidance, authority and control. Although it is still regarded in some quarters as a deft piece of Cold War diplomacy intended to separate the Soviet Union from its fissionable materials,[3] President Eisenhower's speech was generally well received,

2 The Acheson-Lilienthal and Baruch Plans, *US Department of State, Under-Secretary for Public Diplomacy and Public Affairs, Bureau of Public Affairs, Office of the Historian*, 1946. Available online at: http://www.state.gov/r/pa/ho/time/cwr/88100.htm. Accessed June 14, 2009. An earlier and slightly more conciliatory version of the Baruch Plan was developed by nuclear scientist Robert Oppenheimer and proposed under the title of the Acheson-Lilienthal report. Its terms were subsumed by the Baruch Plan.

3 Cf: USC Statement on 50th Anniversary of Eisenhower's Atoms for Peach [*sic*] Speech, December 8, 2003, *Union of Concerned Scientists*, Cambridge, Mass. Available

not least because it carried a message of hope and expectation that the nuclear threat could, in fact, be beaten through collective and cooperative action.

The speech was the opening gambit in a bold, visionary and ultimately naïve policy by the United States to promote the peaceful use of nuclear energy as widely around the "free world" as possible. Ironically enough, one of the enduring legacies of Eisenhower's new Atoms for Peace policy shift and vision was to be the misuse of the nuclear technologies and materials it delivered around the world, in direct contravention of the terms under which the United States had supplied them. By providing the fissionable materials (enriched and *highly* enriched uranium, and plutonium-239) for both power and research reactors, together with direct assistance in building them, the US unwittingly supplied an enhanced potential for broad-based nuclear weapons proliferation in a number of countries, some of which (such as India, Pakistan, South Africa and Israel) went on to produce operable nuclear devices and weapons.

The appearance in 1957 of the International Atomic Energy Agency, operating under the "Atoms for Peace" banner as an autonomous international organization reporting to both the United Nations General Assembly and Security Council, was a true watershed in the development of internationally coordinated nuclear policy and its implementation. A summary of its mission statement reveals its *raisons d'être*:

- an independent, inter-governmental, science and technology-based organization, serving as the global focal point for nuclear cooperation
- assists member states in planning for and disseminating technology and knowledge of the peaceful use of nuclear science and technology
- develops, promotes and maintains high safety standards for nuclear energy, including protection of human health and the environment against ionizing radiation
- verifies by inspection the compliance of states with their commitments under the Nuclear Non-proliferation Treaty and other nuclear agreements to use nuclear facilities only for peaceful purposes.[4]

In the intervening fifty-three years since its constitution, the IAEA has evolved into the world's single most important international nuclear organization, currently with 146 member states and universal recognition of its unique standing and expertise in international nuclear affairs. In essence, the future of nuclear energy is closely tied to the success of the work of this vital international agency. Furthermore, the history of nuclear energy over the past half century has been played out within the context of the international coordination and cooperative programs developed

---

online at: http://www.ucsusa.org/nuclear_power/nuclear_power_risk/nuclear_proliferation_and_terrorism/atoms-for-peace-speech-ucs.html. Accessed June 14, 2009.

4 Mission Statement, *International Atomic Energy Agency,* Vienna, Austria. Available online at: http://www.iaea.org/About/mission.html. Accessed June 16, 2009.

by the IAEA itself, or under its guidance. As nuclear technology is ever more thoroughly diffused through the international community of states, the success of two of the three pillars of the IAEA's vision of nuclear cooperation, namely its nuclear safeguards and verification activities, and safety and security work, become ever more vital.

Since the inception of the IAEA, much of the most promising work on multilateral steps to secure the nuclear fuel cycle has been conducted under its aegis and counsel. As the process and complexities of nuclear proliferation accelerated over half a century, and its consequences began to manifest themselves (in, for example, the Cuban Missile Crisis of 1962 and the 1974 test by India of a "peaceful" nuclear explosive device), so too did new and innovative thinking on the development of practical barriers to nuclear diffusion. However, and as the IAEA knows only too well, asking sovereign states to agree to limit, in any way, their rights of access under the 1970 Nuclear Non-proliferation Treaty to the peaceful uses of nuclear energy (including doing so in the cause of weapons non-proliferation) is often a difficult and fruitless task. Even so, India's 1974 test brought forth a number of regional, multilateral and international initiatives during the nineteen seventies and eighties aimed at reinforcing the NPT's twin goals of discouraging nuclear diffusion, while strengthening the rights of states to the peaceful use of the atom. Examples of these IAEA programs include evaluations of regional and international fuel cycle centers, as well as a scheme for international plutonium storage. For reasons unconnected with their technical viability, such as cost and political opposition, all eventually failed to make progress.

Thus, by 2004, IAEA Director General Dr Mohamed ElBaradei was convinced that the only viable alternative in the search for answers to nuclear proliferation lay in a return to the multilateral nuclear approaches (MNAs) championed in the aftermath of the Second World War and noted above. The 2005 report of the Expert Group appointed by ElBaradei to re-examine the MNA approach marked a clear change of direction for both the IAEA and the wider non-proliferation community. The Group had been asked to provide an overview of the policy, legal, security, economic, institutional, and technological incentives and disincentives for cooperation in multilateral arrangements for the front and back ends of the nuclear fuel cycle. Their reported conclusions and recommendations, delivered in February, 2005, provide a sound and legitimate base for the development of a renewed paradigm in the field of nuclear fuel security, namely multilateral nuclear approaches based on transparency and international collaboration. The five nominated approaches lie along a continuum marked by increasing levels of collaboration, government intervention and innovation.[5]

---

5 Pellaud, B., Nuclear fuel cycle: Which way forward for multilateral approaches?, *IAEA Publications: Bulletin 46/2*. Available online at: http://www.iaea.org/Publications/Magazines/Bulletin/Bull462/nuclear_fcycle.html. Accessed June 18, 2009.

Although these alternative MNAs will be discussed in more detail in Chapter 8: "Old and New Strategies for Non-proliferation Progress," their outlines can be noted here. They can be summarized as:

- Retaining market-based fuel handling arrangements while enhancing the transparency of fuel supplier intent through government backing
- Enlisting IAEA participation as a fuel bank administrator and guarantor of international fuel supply
- Converting existing fuel facilities for use in a multilateral nuclear arrangement
- Creating regional arrangements for new and integrated facilities – "power parks" – based on joint ownership, drawing rights or co-management for all sensitive stages of the nuclear fuel cycle
- Expanding the penultimate scenario to the global level if the nuclear energy industry continues to grow.[6]

North Korea and the Nuclear Fuel Cycle

North Korea, or the Democratic People's Republic of Korea (DPRK), is a closed and isolated state like no other in the modern international system, and exemplifies in its behaviors many of the structural and operational failings of international non-proliferation policy, and its application. Standing in turn defiant and demanding of its South Korean counterpart, its regional neighbors and the Western Alliance generally, North Korea has succeeded in overcoming all impediments thus far placed in its nuclear path. As a result, it has now reached indisputable nuclear status with the underground detonation of small nuclear explosive devices in October, 2006 and May, 2009. When added to its steady progress in the development of short and longer range missile technology, together with its demonstrated willingness to export nuclear and missile technology and materials, North Korea presents a regional and global diplomatic challenge of enormous proportions. Even the bare outlines of North Korea's involvement in the nuclear fuel cycle shows the nature and extent of the failings of the nuclear non-proliferation regime.

North Korea has demonstrated an interest in nuclear affairs since before the Armistice which halted Korean War hostilities in July, 1953. In 1956 it demonstrated the level of that interest by beginning to send nuclear specialists for training in the Soviet Union. During the 1960s it initiated its own basic nuclear research program, and its technicians received training in nuclear fuel reprocessing and plutonium separation techniques at Soviet nuclear reprocessing plants. In 1961, North Korea began to build its Yongbyon nuclear research complex, while cementing its alliance with the Peoples Republic of China by signing a Treaty of Friendship, Cooperation and Mutual Assistance with its giant northern neighbor.

6 Ibid.

By 1963 it had procured, and was installing, an IRT-2000 Soviet-supplied nuclear research reactor with the capacity to use 80 per cent enriched uranium as fuel. The following year, the North Korean government in Pyongyang was claiming the discovery, with assistance from China, of some four million tons of commercial grade uranium ore, while its Yongbyon complex had been completed.[7]

Clearly, then, North Korea consistently demonstrated its determination to become a nuclear state almost from the moment of its appearance on the world stage. In 1967, North Korea's "Great Leader" Kim Il Sung had ordered the development of nuclear warheads for future missiles, as the 2 MW (th) IRT-2000 research reactor had become operational. By the mid-1970s, North Korea had embarked on nuclear weapons and nuclear fuel-related research and development and had, surprisingly, joined the International Atomic Energy Agency. This latter move was later interpreted in South Korean non-governmental circles as a ploy to gain access to the wealth of materials and data available within the technical resources of the Agency. It was also during this period that North Korea began small scale pilot operations to separate plutonium from spent fuel extracted from its research reactor at Yongbyon.[8]

This summary of North Korea's nuclear involvement up to the mid-1970s provides ample evidence of some of the most intractable challenges posed by the structures and processes of the nuclear non-proliferation regime. What this chronicle of effort shows is a consistency of purpose, starting from a very low baseline position, in the face of (admittedly early and weak) measures from the international community to verify and enforce its adherence to the spirit and *raison d'être* of the 1970 Nuclear Non-proliferation Treaty, a multilateral agreement *in international law* to which the North Korean state acceded in 1985, and purported to leave in 2003.

What it exposes above all, however, is the structural and procedural failure of international law and international diplomacy of that era to rein in what was already obvious by the mid-1970s, if not long before (not to mention the inability of democratic governments to enforce private sector export controls over nuclear-related materials and technologies). This amounted to nothing less than the clear, long-term intention of the North Korean leadership, vigorously and consistently pursued, to equip the state with the capacity to produce deliverable nuclear weapons. Along the way, North Korea has also successfully undertaken the export of both nuclear and ballistic missile technologies.

The central issue inherent in this story of non-proliferation failure is whether (and if so, how) any state similar to North Korea on relevant criteria could, in the modern era, emulate the feat of the DPRK and produce crude nuclear weapons using an indigenously developed nuclear fuel cycle as a source of fissionable

---

7 North Korea profile, Nuclear chronology 1942–1989, *Nuclear Threat Initiative*, Washington, DC. Available online at: http://www.nti.org/e_research/profiles/NK/Nuclear/46.html. Accessed June 20, 2009.

8 Ibid.

material and technical expertise. Having done so, what then would prevent their acquisition by terrorist groups bent on the most destructive attacks imaginable on the great cities of the West, if such an aim were endorsed by the state in question? The major factors operating at the opening of the new century both to inhibit, and fail to inhibit, such an outcome in a calculus of the probabilities of nuclear non-proliferation success and failure can be condensed in the following way:

Factors inhibiting fuel cycle proliferation

- Slow but discernible moves within nuclear states reliant on others for nuclear fuel cycle services towards acceptance of the need for agreed limits to full direct fuel cycle participation, on the basis of assured access to nuclear fuel supplies
- continuing US funding of the removal for down-blending in the United States of remaining weapons-usable highly enriched uranium from research reactors and stockpiles worldwide
- continuing efforts by the United States and within the UN to secure legacy fissionable materials within the former Soviet Union
- a more cooperative global approach to the prevention and interdiction of the illicit distribution of nuclear technologies and materials
- the increasing scrutiny of nuclear security created by the growing realization of the future importance of nuclear power generation to the world's natural environment
- the emergence of new and more proliferation-resistant reactor and reprocessing technologies and systems.

Factors failing to inhibit fuel cycle proliferation

- The diminishing, yet still substantial, quantity of highly enriched uranium in use or storage in connection with nuclear research reactors around the world. Much of this is inadequately secured against theft or diversion to weapons development programs
- the development of civilian plutonium production reactors in China and India which enhance the future use and diffusion of plutonium fuels as a substitute for enriched uranium
- failure to make progress on a multilateral treaty to mandate the cessation of weapons-grade fissile materials production
- incomplete or *ad hoc* approaches to national and multilateral oversight and control of the commercial trade in dual-use nuclear technologies
- insufficient attention to the need to stress to developing and aspirant nuclear states the energy security benefits available from internationally conducted programs guaranteeing nuclear fuel services
- the lack of universal agreement on defining effective global nuclear security, control and accounting standards for nuclear stockpiles, including physical

security standards (despite legal obligations on all states to adhere to such standards under UN Security Council Resolution 1540).

North Korea's more recent nuclear history brings up to date a consistent pattern of action by the North Korean state in pursuit of recognition by the rest of the world, but principally by the United States, of its status as a nuclear power. Notwithstanding the long story of North Korea's checkered diplomatic struggles with neighboring powers Russia, China, Japan, South Korea and the United States since 2003, when the DPRK withdrew from the NPT, the leadership of this enigmatic dictatorship appears to have driven a hole straight through the edifice of the nuclear non-proliferation regime. In doing so, it has certainly shone a harsh light of failure on this most vital of international security architectures.

Can lessons be learned from this experience by the leaders of powerful nations and their policy advisers, or should this singular episode be weighed against similar examples of challenge to the paradigm of non-proliferation? How much international regime building remains to be accomplished before the current imminent threat of a nuclear weapons break-out is, at the very least, reduced to manageable proportions? We now turn to consider the nuclear history of India to construct a response to these questions.

India and the Nuclear Fuel Cycle

India's engagement with the nuclear fuel cycle has, on the face of things, been a very different experience from that of North Korea. Whereas North Korea's drive towards a full nuclear fuel cycle was motivated from the beginning by the need felt in Pyongyang for the acquisition of nuclear weapons, India's interest in nuclear affairs began as an investigation into the possibility of using her abundant supply of thorium (approximately one quarter of the world's high quality deposits) to generate cheap nuclear energy, following independence from Britain in 1947.

Once attained, however, India then used its newly acquired nuclear expertise to develop nuclear explosives in the aftermath of its defeat by China in the border war of 1962.[9] India duly demonstrated its nuclear expertise by detonating its only "Peaceful Nuclear Explosion" (PNE), an underground test of a non-militarized device, in May, 1974, a full decade after China's first test of a fission device in 1964. India finally undertook the militarization of its nuclear explosives technology under the government of Prime Minister Rajiv Ghandi during the late 1980s, in the face of growing concern over rival Pakistan's own burgeoning nuclear weapon development program.

---

9 India profile, Nuclear overview, Historical overview, *Nuclear threat initiative*, Washington, DC. Available online at: http://www.nti.org/e_research/profiles/India/Nuclear/. Accessed June 23, 2009.

In the years following, India has steadily expanded its investment in both its civil and military nuclear programs. On the military side, India has built what it terms a "credible minimum deterrent" nuclear posture on the basis of the two rounds of test explosions it conducted in May, 1998, having been in possession of operational (but untested) nuclear weapons for delivery by aircraft or missile since 1994. With a formal, credible and civilian-led command, control and communications system for its newly acquired deterrent force and, importantly, one that is increasingly resistant to terrorist attack, India has declared its nuclear doctrine to be one of "no first use" or "retaliation only," and then only against states which possess nuclear weapons or are aligned with a nuclear weapon state[10]. As a result of these developments, India is now unequivocally a member of the nuclear weapon state community, reportedly now with around fifty fully operational nuclear weapons, but one which has not acceded to the Nuclear Non-proliferation Treaty and remains committed to its refusal to join what it regards as a deeply flawed and fundamentally inequitable international bargain.

The consequences for India of its consistent refusal to join the NPT have been far reaching for its aspirations to develop a viable civilian nuclear energy program. As a long standing non-member of the Nuclear Suppliers Group, and without either a legitimate place within the international uranium trading system, or adequate indigenous uranium deposits, India has had to fall back on its own resources. Thus, long term nuclear research in India has for many years been centered on the prospect of constructing an industrially developed economy built with the assistance of thorium energy production and advanced plutonium breeder reactor technology. Even so, India's long-established research efforts aimed at developing a viable thorium fuel cycle, and in the field of fast neutron breeder reactors, are yet to bear fruit, not least for a raft of technical reasons. Its choice for a feasible nuclear power generation industry has therefore been limited to 'conventional' uranium-based nuclear energy production which, in the face of severe uranium supply problems, still accounts for only three per cent of India's total electricity demand.[11]

Notwithstanding these obstacles to progress, India has, nevertheless, quietly and consistently built up significant nuclear-related technical resources and skills out of sheer necessity, a process that has resulted in a far smaller degree of reliance today on international suppliers of nuclear technology and expertise than would otherwise have been the case. In the long term, India's massive commitment of resources to nuclear research and development over many years has been rewarded with significant progress. This was greatly assisted in its early stages by the decision of the government of Canada in 1963 to supply India with its Canadian Deuterium – Uranium (CANDU) heavy water/natural uranium production and research

---

10 Ibid.

11 Gadekar, S., India's nuclear fuel shortage, *Bulletin of the Atomic Scientists*, August 6, 2008. Available online at: http://www.thebulletin.org/web-edition/features/indias-nuclear-fuel-shortage. Accessed June 23, 2009.

reactor technology. Although the plutonium used in the 1974 PNE explosion was not the product of a CANDU production reactor (it had been generated from 1960 by the CIRUS Canadian-supplied "Atoms for Peace" program research reactor using CANDU technology) the reality of Canada's transfer of its innovative nuclear technology was instrumental in the subsequent need to train a new cohort of nuclear scientists and technicians for India's civilian nuclear energy program. This in turn led the way for their later employment in a weapons development program. It was perhaps not a coincidence that the CANDU technology sought and received by India allowed its research reactor to be refueled, and spent fuel containing plutonium to be removed, without the need for it to be shut down, thus disguising the schedule by which spent fuel would be reprocessed to extract weapons-grade plutonium.

Other effects and consequences of India's nuclear proliferation activities also gradually became apparent. First, the Nuclear Suppliers Group (NSG), a grouping of states with interests in the manufacture and export of technologies with both peaceful and military nuclear applications, would not otherwise have been constituted at the time, nor in the manner, in which it actually appeared. In fact, the NSG and its mission were established as a vital institutional element of the early nuclear non-proliferation regime as a *direct response* to India's 1974 PNE explosion. Since 1974, the NSG has evolved into its current role of providing Guidelines and Trigger Lists to its member states in relation to the transfer of nuclear material, equipment and technologies used in the nuclear fuel cycle, as well as items with legitimate dual uses both within and outside nuclear power generation. In the case of India's nuclear agreement with the United States of October, 2008 (discussed below) one immediate effect was the NSG's issue of a waiver of the ban it had imposed on all member states in respect of weapons-related and dual use nuclear trade with India.

A second aspect of India's pursuit of nuclear weapons arises from the processing plant used to extract from the CIRUS research reactor's spent fuel the plutonium subsequently employed in the 1974 test explosion. Although claimed by India and the US as an indigenous facility, the plant had, surprisingly enough, been constructed by a secret consortium of US and European corporations. Furthermore, the United States had secretly supplied the heavy water required to make CIRUS operable.[12] Eventually, a committee of the United States Senate became aware of these activities, and drafted what later became the Nuclear Non-proliferation Act (NNA) of 1978.

This outcome is important because the NNA, in response to the CIRUS experience, immensely strengthened the effectiveness of US participation in the

---

12 Leventhal, P., CIRUS reactor's role in a US–India Nuclear Agreement, Presentation to Center for Nonproliferation Studies, *Nuclear Control Institute*, December 19, 2005. Available online at: http://www.nci.org/06nci/04/CIRUS per cent20Reactorsper cent20Role per cent20in per cent20a per cent20US-India per cent20Nuclear.htm. Accessed June 26, 2009.

global regime of non-proliferation by mandating full-scope IAEA inspection safeguards against the diversion of US-supplied nuclear materials, equipment and technology from peaceful to belligerent use. It also sought to inhibit worldwide spent fuel reprocessing, which had produced India's first plutonium bomb material (as well as national control world wide over the front end of the nuclear fuel cycle) through an International Nuclear Fuel Authority. Such an Authority would be complemented by an internationally managed fuel bank, as well as uranium enrichment and spent fuel repository facilities.

In these goals, which appeared well before any hint of widespread and positive political will towards the fuel bank idea existed, the NNA was spectacularly unsuccessful, and remains so today. Nevertheless, the multilateral nuclear control aspirations outlined in the US Nuclear Non-proliferation Act did establish at least the framework of a future internationally controlled nuclear fuel cycle. As we shall see in later chapters, multinational nuclear agreements of these kinds are now being taken very seriously indeed.

As for India's plutonium-producing CIRUS research reactor, it had been supplied by Canada on the basis of "peaceful use only" contractual arrangements under Canadian and US jurisdictions, and under the terms of the US Atoms for Peace nuclear technology dissemination program. The vulnerability of that Eisenhower era program to misuse was, of course, plain for all to see in the form of India's initial "peaceful" nuclear test of 1974. The entire episode still provides a salutary lesson in the growing propensity of states – including those with democratic political systems such as India – to fail to resist the temptation to acquire what its political leadership believes is an essential element of national security. This is especially so in the case of nuclear weapons technologies *just because of* the massively destructive nature of nuclear explosions, rather than in spite of it. The case of the Indian nuclear weapon development program is a prime example of this truism.

A final part of the Indian nuclear story with relevance to nuclear fuel security, and with resonance in today's world of fluid and dynamic nuclear proliferation challenges, is the controversial US–India Nuclear Agreement of 2008. Whatever the alleged rights and wrongs, benefits and drawbacks of the deal, there is no doubt that it represents a new direction in global non-proliferation policy and practice. Announced by President Bush and Indian Prime Minister Manmohan Singh on July 18, 2005 and given final legislative approval by the United States Congress on October 1, 2008, the Agreement frees up nuclear and nuclear-related trade between the two countries which had been frozen for thirty years (since the 1974 PNE explosion). It can also be noted that India remains outside the Nuclear Non-proliferation Treaty, its resistance to which is unaffected by its partial admission to the international nuclear fuel cycle which this Agreement represents.

In essence, the new Agreement provides substantial US assistance to India's civilian nuclear industry, as well as expanding cooperation between the two countries in energy and satellite technologies. At first glance, the deal appears to contain the essential ingredients of any successful contract or diplomatic meeting of minds: mutual, balanced and realizable benefits. In exchange for a number of concessions which effectively bring India within the aegis of a number of global nuclear non-proliferation principles and practices, India has been granted access to much-needed uranium fuel for its civil production reactors from the United States (and from all other Nuclear Supplier Group countries). India is also free to order and commission production reactors and associated technologies, including dual use technologies suitable for nuclear weapons manufacture, from the same sources.

On the other side of the equation, India has accepted full-scope (essentially comprehensive, intrusive and short notice) safeguards inspections by the Inspectorate of the IAEA of fourteen of its twenty-two civil power reactors, verifying that no diversion of materials, technologies or equipment to a military program has taken place at those sites. Furthermore, all future power reactors – declared as such solely by New Delhi – will be subject to the IAEA full scope safeguards regime. However, military facilities and stockpile will not be subject to any safeguards or inspections of any type.

Importantly, though, India will maintain its moratorium on nuclear weapons testing, introduced following its multiple tests performed in 1998, while it further strengthens the security of its nuclear arsenal. Finally, India has undertaken to work towards the negotiation of a multilateral Fissile Materials Cut-Off Treaty, and act in support of nuclear non-proliferation initiatives generally, while agreeing to help prevent the spread of uranium enrichment and spent fuel reprocessing technologies to states not in possession of them.[13]

It is not unrealistic to claim that the US–India Nuclear Agreement of 2008 is one of the most significant developments in nuclear non-proliferation in recent years. Those opposing the deal have described it as too biased in favor of Indian interests. For example, Henry Sokolski, Executive Director of the Nonproliferation Policy Education Center, has been quoted as pointing to the inadequacy of safeguards against the Indian government continuing to enhance its nuclear arsenal:

> We are going to be sending, or allowing others to send, fresh fuel to India – including yellowcake and lightly enriched uranium – that will free up India's domestic resources of fuel to be solely dedicated to making many more bombs than they would otherwise have been able to make.[14]

Other objections from the US non-proliferation community include the fact that these arrangements do not cover fissile material produced by India prior to the Agreement, nor place any cap on future production of fissile material for military applications, as established nuclear weapon states are increasingly moving

13 Backgrounder: The US–India nuclear deal, *Council on Foreign Relations*, October 2, 2008. Available online at: http://www.cfr.org/publication/9663/#1. Accessed June 26, 2009.

14 Ibid.

towards. Furthermore, there is no mandated restriction on the number of nuclear weapons India may produce.

More positively, the deal has found praise as a step change in US policy approaches in the field of intractable or difficult nuclear proliferation issues. By extension, this judgment is also valid in terms of its influence on global non-proliferation policy evolution. For example, although India is an NPT recalcitrant, the deal demonstrates that this need not preclude its recognition as a responsible steward of nuclear power and technological innovation. In pursuit of commonly held regional strategic goals such as countering the rising influence of China, the Agreement engages pragmatically with India on nuclear affairs which, it asserts, do not pose a proven proliferation risk. Rather than punish India for its nuclear policies, especially in respect of membership of the NPT, the Agreement rewards its restraint, not least in its record of abiding by the strict guidelines imposed on member states by the Nuclear Suppliers Group. In other words, the Agreement provides implicit recognition that there is more to be gained from working with India than against it.

Such a conclusion should stand as a guide for any decision maker or government adviser in the field of nuclear fuel security and the wider arena of global nuclear non-proliferation policy and action.

Pakistan and the Nuclear Fuel Cycle

The story of Pakistan and the nuclear fuel cycle is a thematic thread linking much of this study's analysis of global nuclear fuel security. The nuclear affairs of Pakistan clearly demonstrate well-defined divergence from the Indian example in terms of their genesis, evolution, proliferation risk profile and strategic security implications. Whereas India is, after a long journey, finally gaining acceptance in the international community as a responsible nuclear state, Pakistan's history of the use of illicit international nuclear smuggling and the export of nuclear-related technologies, for whatever purposes, clear distinguishes it from India in the nuclear field.

The agency of individuals in influencing the engagement of the Pakistani state in global nuclear affairs was introduced in Chapter 4 when discussing the A.Q. Khan nuclear smuggling network. The astounding story of Khan's ability to work simultaneously both within and beyond the oversight and control of the state in the process of providing, and then offering for international sale, his country's nuclear technologies over many years is emblematic of the scale of distinction between Pakistan and its larger rival. The saga of the Khan network's activities, the extent and nature of its operations, and their ultimate effects has a number of years to run before final conclusions are determined, consequences evaluated, and policy responses refined – in A.Q. Khan's homeland, its region and globally. It may be some time, in fact, before there can be complete confidence that Khan's shadowy influence has been irrevocably removed from the world of illicit international nuclear activities. Even so, and of more direct and contemporary concern for the United States, its allies and beyond, lies Pakistan's place at the crossroads of nuclear weapons and terrorism.

Pakistan has acquired nuclear weapons as a deterrent and strategic counter-move in reaction to its fear that India's growing and modernizing nuclear program would eventually result in its capacity to launch a non-survivable nuclear attack in the event of full scale war between Pakistan and India. As a result, Pakistan now possesses, on some estimates, around 85 operational nuclear weapons, to which it continues to add further units. Recalling that Pakistan, like India, has not signed the NPT, it is significant to note that in October, 2008, China agreed to supply Pakistan with two new nuclear power plants, a development that the 2008 Report to President Bush of the Commission on the Prevention of Weapons of Mass Destruction Proliferation and Terrorism interpreted as signifying "a nascent nuclear arms race in Asia."[15]

It is a measure of the Pakistan-based terrorist nuclear threat as perceived in Washington that its importance is placed by this Report, which was called for by Congressional legislative action, on an equal footing with the nuclear deterrent policy position of the Pakistani state. Equally disconcerting for nuclear non-proliferation planners is the fact that strategic thinking and planning in Pakistan is overwhelmingly focused on the supposed existential threat posed by a nuclear-armed India, a reality which distracts government attention away from the need to balance this threat with that of the theft of nuclear weapons by terrorist organizations, and its nuclear proliferation implications.

Pakistan's nuclear weapons are under the operational control of its military establishment, which asserts their complete immunity from theft by al-Qaeda or Taliban Islamic extremist groups. Nevertheless, should the Pakistani nuclear arsenal grow as the state becomes weaker, the security of that arsenal will be called increasingly into question in Washington, in the South Asian region, and more broadly by the international community. Terrorist safe havens in Pakistan's Federally Administered Tribal Areas (FATA) on the border with Afghanistan must, in the Report's view, be a focus of counter-terrorist operations for the United States and the Pakistan Government. This conclusion is reinforced by the 2007 National Intelligence Estimate of the Director of National Intelligence, which determined that al-Qaeda has "protected or regenerated key elements of its homeland attack capability, including a safe haven in Pakistan's Federally Administered Tribal Areas." The Estimate further asserted that al-Qaeda will continue to try to acquire, and then without hesitation use, WMD materials, including those with radiological and nuclear implications.[16]

15 *World at Risk: Report of the Commission on the Prevention of WMD Proliferation and Terrorism*, 2008, New York, NY, Vintage Books. At p. 67. Available online at: http://www.preventwmd.gov/report/. Accessed July 1, 2009.

16 Ibid. At pp. 65–68.

It is clear, then, that Pakistan's nuclear situation presents the world with a dangerous microcosm and vision of an even more lethal, and indeed global, possibility: accelerating nuclear proliferation driving credible threats of nuclear attacks by terrorist organizations, often aided by illegal international trade in military and dual-use nuclear technologies, and direct assistance from some nation states. Furthermore, a descent into regional nuclear arms races, in a classic process of threat, counter-threat and misperception, is limned in sharp relief by even a casual reading of the record of nuclear non-proliferation success and failure among the three distinctive nuclear states discussed in this chapter.

While the North Korea and Pakistan cases provide cause for serious doubt about the prospects for renewal and regeneration of resistance to global nuclear diffusion, the US–India case allows some room for optimism that progress through policy accommodation, engagement and new flexibilities is at least *possible* on a global scale. Much will depend on the generation of sustainable levels of trust in the possibility of lasting international proliferation oversight and control offered by new multilateral nuclear agreements (MNAs). While the notion of feasible MNAs is not new, as US non-proliferation legislative history shows, the rapidly accelerating spread of nuclear weapons capacities, with its accompanying and growing nuclear terrorism threat, certainly is.

Another prospect is offered by the success of new proliferation-resistant nuclear fuel technologies, such as the new Generation IV class of production reactors, whose designs are based on denial of access to weapon-usable fissionable materials through the lack of back end separation processes. Such processes are themselves believed to be well beyond the capabilities of all known terrorist organizations.

This study further discusses the role of new nuclear technologies in later chapters, while the following chapter engages in a wide ranging examination of the agency of enriched uranium in helping to shape the global strategic landscape and, by extension, homeland security. The most telling factor, which will be examined in depth, is the "pull" of the nuclear fuel cycle's capacity to produce the enriched uranium and plutonium which makes feasible a clandestine nuclear weapon development program. For some states whose governments believe that there is, on balance, more strategic or regional advantage to be gained than lost from such a program, the temptation to divert and enrich uranium for use in a weapon program may become too great to ignore.

Iran will be used to illustrate this general point, and will also be examined in terms of its putative motives in setting up large scale uranium enrichment operations. The broader strategic implications of Iran's nuclear activities for the United States and its allies, as well as the Middle East region, Western and Eastern Europe, and Russia will illustrate the regional and strategic difficulties which are embodied in externalities of this kind when they intersect complex security systems.

Chapter 6
Strategic Security and Enriched Uranium

Introduction

Dean Acheson, Secretary of State from 1949 to 1953 in the Truman Administration, played a central role in defining the essence of US Cold War strategic policy. When asked once to define the term "foreign policy," his response, after considerable pause for thought, was "Foreign policy is one damn thing after another".[1]

From the perspective of an individual operating at the zenith of US foreign and strategic policy formulation and practice, including global security policy, this is a reasonable opinion to hold. From the point of view of those attempting to make sense of the sweep and flow of international events over time, and without the benefits afforded by high government office, such a conclusion needs to be tempered by an acknowledgement that there are, in fact, discernible and replicable patterns in the affairs of nation states. Those patterns, condensed to form a repeatable process of cause and effect, stimulus and reaction, often prove useful as sign posts for those seeking effective policy change in response to security challenges. Here, another of Acheson's famous aphorisms, "Always remember that the future comes one day at a time," is also relevant as a pithy yet apposite reminder that small and incremental movements which, of themselves, do not materially increase a risk or require an immediate alteration to national policy, each contribute in a small but significant way towards an undesirable, even threatening, final outcome. A failure to acknowledge this reality, and to respond in a timely manner to incremental change, carries with it – all else being equal – the possibility of systemic policy failure of the kind which now threatens to de-rail the global structure of nuclear non-proliferation policies and their implementation.

So it is with the specific issues and questions, canvassed throughout this discussion, surrounding nuclear fuel security and its place within the wider challenges presented by the global proliferation of nuclear weapons, their fissile materials and technologies. The revelation in 1987, for example, that the security services of Saddam Hussein's Iraq had used "front" companies as false end-users for the importation from the United Kingdom of highly sensitive and internationally controlled components used in the construction of uranium enrichment centrifuges, is one more brick in the wall of resistance against weapons proliferation. The detailed intelligence obtained by the IAEA on how this was accomplished,

1 Nunn, S., Speech to American Nuclear Society Annual Meeting, June 15, 2009. Available online at: http://www.nti.org/c_press/speech_Nunn_American_Nuclear_Society061509.pdf. Accessed July 4, 2009.

including effective Iraqi control of the UK manufacturer involved – Matrix Churchill – amounts to a valuable addition to the sum of multilateral knowledge about the shape and capabilities of well-funded and determined sovereign state nuclear proliferators.[2] As always, knowledge is power.

There remain, unfortunately, circumstances in which the governments of states (such as Colonel Mu'ammar al-Gaddafi's Libya) believe that the balance of penalties and benefits available from the development of nuclear weapons capacities and potentials is weighted on the positive side in terms of national security and national interests. When this judgment is accompanied by clear opportunities to test the truth of such a calculus, it is inevitable that the temptation to do so can, in some circumstances, overcome caution and become irresistible. The existence in such a mix of a strongly totalitarian regime will, of course, only strengthen the probability of that state testing the limits of the international community's patience.

The case of Iran's nuclear program will be used in this chapter to illustrate this general point, and will also be examined in respect of the motives imputable to the decisions and actions of its government in establishing large-scale uranium enrichment operations, principally at its nuclear facility at Natanz. Central to these considerations are, among others, Iran's importance as a major regional power in the Middle East, together with the growing missile threat that Iran is now perceived by European states to pose to their national security. Iran's expected move towards nuclear weapons manufacture and weaponization, added to its demonstrated and maturing ballistic missile development capacities, was a pivotal factor in the US-led proposal to build a missile defense shield in the Czech Republic (radar facilities) and Poland (missile interceptor batteries). Israel's reaction to these developments further complicates the regional strategic calculus.

The response of an increasingly confident Russia to Iran's controversial nuclear policies, as well as what Moscow believes to be a pattern of steady encroachment into Eastern Europe from the west by the North Atlantic Treaty Organization (NATO) lends a further layer of strategic uncertainty. Even so, and as discussed earlier, Russia has itself willingly participated in an extended process by which much of its discarded weapons grade HEU has been removed from Russian soil and converted through down-blending for use in American nuclear power plants.

The most pressing strategic security policy questions which arise from the illicit or covert diversion or conversion of fissile material for use in weapons development programs are based on the urgent need to bring all available nuclear fuel non-proliferation strategies to bear on states which perceive substantial benefit in their acquisition of nuclear weapons. The three discrete fields in which such efforts must now be concentrated are in:

---

2 Nuclear black markets: Pakistan, A.Q. Khan, and the rise of proliferation networks, Chapter 2: Nuclear Black Markets: Other Countries and Networks, *International Institute for Strategic Studies*, 2008, London, UK. At pp. 45–6. Available online at: http://www.iiss.org/publications/strategic-dossiers/nbm/. Accessed July 6, 2009.

- verification and enforcement of the NPT-based prohibition against transfer of nuclear materials from energy production to weapons programs
- enhancement of the physical security of legitimately held nuclear materials and technologies
- further progress in strengthening the security of nuclear fuels in both the front end (HEU enrichment) and back end (spent fuel reprocessing and plutonium extraction) of the nuclear fuel cycle.

This chapter will discuss globally important strategic issues germane to these non-proliferation imperatives against a background marked by the effects of the unique properties of HEU and plutonium on the strategic security calculations of states, both individually and as members of international alliances and organizations.

Strategic Security

Strategic security is a concept which is embedded in the national interests of nation states in ways which demand careful attention by nuclear policy planners and scholars. When the international context and specific effects of states' combined efforts to maintain and defend their strategic security are not fully comprehended by all interested parties, the result can be misleading and dangerous (mis)understanding of both capacities and intent, followed by rising international tensions and possible conflict. When such a situation involves states armed with nuclear weapons, or with an intention to acquire them, the result may be as inadvertent as it is catastrophic. The increasing possibility that the political decision makers of states will see the use of nuclear explosives as the *only* answer to a perceived threat, or – ultimately even worse – a realistic *alternative* among other, less drastic, responses, demands close engagement with the primary non-proliferation instrument of nuclear fuel security in ensuring that such a policy choice is never made under any circumstances.

Serious attention to the contemporary nature of international security dates only from the early 1980s, when a group of scholars, epitomized and led by individuals such as the international security theoretician Barry Buzan, gained increasing recognition from that time by questioning prevailing security orthodoxies.[3] A context for this apparent heresy can be found in a shift in emphasis from "national security" to "international security" in the search for greater global military stability, a phenomenon which had been a hallmark of the later Cold War era. Furthermore, this change of emphasis in conceptualizing security was distinguished by an acknowledgment that it ought to be possible for a state to enhance its own security without necessarily (and unnecessarily) diminishing that

3 Note, for example: Buzan, B., 1983, *People, states, and fear*, Brighton, UK, Harvester Wheatsheaf; and Buzan, B., Waever, O., and De Wilde, J., 1997, *Security: A new framework for analysis*, Boulder, Lynne Rienner.

of its neighbors. This amounted to a strengthening belief that international politics need not be, at base, a zero-sum game of winners and losers.

Later still, as the 1990s arrived, the very concept of "security" was undergoing a searching reappraisal as the traditional realism which underpins the foundations of mutually assured nuclear destruction as a *plausible* defensive strategy was increasingly seen as conceptually and politically bankrupt. Furthermore, as new and urgent problems continued to build throughout the post-Cold War world – in many fields of government policy from environmental degradation to demographic change and human rights claims – the need to greatly broaden the constituent field of "security" was also acknowledged as overdue. In essence, the new security paradigm, first developed and expressed within Buzan's Copenhagen School of critical international security scholarship, was based on a claim that what ultimately *counts* as a security issue evolves from political and social discourse, at least in more or less democratic political systems, rather than exclusively from popularly representative elites and other change agents with influence and status.

This is not necessarily to hold a view that a realist perspective on global relations no longer provides the starting point for the development of any state's international policy settings, or of their adjustment to changed circumstances. It does, however, allow the entry into strategic thinking of more complexly contingent security considerations, together with a more clearly drawn image of the *depth and breadth* of just what must be secured by leaders of states in order for the nation to survive and, indeed, prosper.

What becomes, then, of the need to make progress on nuclear security, and specifically global nuclear fuel security policy, in a world of increasingly diffuse and open-ended "security" considerations of the kind outlined here? It seems that the most cogent explanation is contained, first and foremost, within the context of an undisturbed traditional concept of national (or homeland) security, based on the primary requirement of government to secure the nation from external and internal subversion and attack. Without doubt, many newer interpretations of national security, not limited to the Copenhagen School example above, must now find a place beside security considerations which are overtly militarily-based, whether these new fields are economic, environmental, social or political in their discrete nature. However, the highest priority for the leaders of all states, and with the greatest claim over the expenditure of national resources is, and will remain, national military security: the capacity to repel any challenge to the territorial integrity and political independence of the state at all times. All else in the field of government and public affairs is contingent upon this single imperative, with the result that the homeland security of the United States and its allies is founded on this premise.

Furthermore, in the context of progress in international fuel security enhancement, there must be a renewed preparedness by governments to work more closely together in order to give greater effect to the reality and consequences of "international security" (in contrast to purely "national security") which is truly global, truly inter-dependent, and hence truly vulnerable to attack upon its integrity in any number of ways, and by a wide range of malicious agents based anywhere around the world. A significant attack on energy infrastructure of any type – for example, on oil production platforms or pipelines – and wherever it occurs, has an immediate effect on world energy prices as well as regional and global economic outcomes. In this conventional sense, the security of nuclear energy assets is no different from any other form of energy infrastructure although, of course, the nature of nuclear fuel and technologies mandate unique approaches to rendering fissile materials immune from illicit diversion, theft and misuse.

The strategic energy security paradigm has, for the reasons noted above, served to strengthen the reach and capacities of what Bull has described as "international society," a concept which he distinguishes from a system of states. The distinction is important in this discussion because it is the transformation of a *system* of states into a *society* of those states which is crucial in the global quest to tackle the world's most pressing questions, from climate change to nuclear security. Put another way, a situation in which states are in regular contact with each other, and in which their interaction is sufficient to demand that the behavior of each must be regarded as a necessary element in the policy calculations of the others, comprises a *system* of (independent) sovereign nation states. They are bound together by economic and strategic imperatives not by choice, but of necessity.

In contrast, and as Bull points out, a *society* of states seeks to enhance relations between its component states through a consciously constructed social contract in which shared values, principles and interests are upheld and reinforced. They accomplish this difficult feat through sharing in the operation of common institutions to enforce a common set of rules, thereby forging orderly, predictable and reliable mutual relationships.[4] In the modern era, the only relevant, though flawed, paradigmatic example on the global scale, it may be claimed, is represented by the United Nations Organization, divided as it is between myriad purposes and programs from food security to military intervention. Only by recognizing this unique global system (not yet society) of states as an indispensable bulwark against the growth of nuclear capacities *as a form of national defense* can the United Nations successfully support the transformation of the single global nuclear watchdog, the International Atomic Energy Agency, from an underfunded agency with a restricted remit to the embodiment of international society which seeks full and enduring nuclear security.

In the nuclear field, then, there must be a goal-driven and global *society* of states, enforcing non-proliferation rules using common institutions which are adequately supported financially, technically and politically by all nation states. In other words, there is a pressing need for global agreement to quarantine the field of nuclear security from all other expressions of international relations for the

4 Watson, A., (1987) Hedley Bull, states systems and international societies, *Review of International Studies*, 13: 147–53. Available online at: http://www.jstor.org/pss/20097109?cookieSet=1. Accessed July 16, 2009.

common good of humanity. There are obvious tensions inherent in the construction of a functionally distinct international architecture of this kind, especially one whose purpose is to thwart the ambitions of nuclear proliferators without rules or scruples.

Can nation states ever successfully isolate the misuse of nuclear fuels, technologies, materials and skills within the context of the full sweep and contingencies of their international and security policies? Should they eventually succeed in doing so, can they lock in the durability of such policy bifurcation in a rapidly evolving international security environment? This is, of course, what the NPT has quite successfully mandated since 1970 on the basis of the nuclear bargain between the five Nuclear Weapon States and all other states. Notwithstanding the failure of the "haves" to fulfill their promise to divest themselves of nuclear explosives, the bargain has, up to this point, substantially survived a number of defections by nuclear-aspirant states (Libya, North Korea, South Africa and, perhaps, Iran). The propensity of additional states to emulate these four, in search of illusive and exclusively nationally-focused strategic security, must be addressed and surmounted.

Strategic Security and Nuclear Fuel

The revelation that a country not previously known to have nuclear ambitions has been implicated in efforts illicitly to acquire some form of nuclear material, equipment or expertise is usually met with widespread shock and consternation, followed by international condemnation and moves to quarantine that state from further access to nuclear-related assets. Libya is an exemplar in terms of the history of its clandestine nuclear weapons development program, which lasted from the early 1980s until the discovery of that program in 2003, and its dismantlement during the following year by the United States and the United Kingdom (with oversight by the IAEA).

The world's only institutional center of nuclear cooperation and surveillance, the Vienna-based nuclear watchdog had failed, over a period of more than twenty years, to detect a multi-faceted nuclear weapons program which included undeclared items such as gas centrifuges (supplied by the A.Q. Khan network), technical design data, 1,200 tons of Niger-supplied uranium ore concentrate and a blueprint, of Chinese origin, for fission weapon fabrication[5]. It could be no consolation for the IAEA and its Inspectorate that the Libyan Arab Jamahiriya had been a member state of the 1970 Nuclear Non-proliferation Treaty since 1975, and had been subject to the IAEA's original inspections regime since 1974.

---

5 This weapon blueprint was supplied by the Pakistani A.Q. Khan nuclear smuggling network. Nuclear Threat Initiative Libya Profile, *Nuclear overview 2007*. Available online at: http://www.nti.org/e_research/profiles/libya/3939.html. Accessed July 16, 2009.

An inspection and verification failure of this kind had the potential to seriously destabilize both regional and global international security conditions. Only through the rapid dismantlement of Libya's nascent weapons program, and coordinated international diplomacy, was rapprochement with the West undertaken and the worst kind of regional security instability avoided. The Libyan case is a welcome but increasingly rare example of a country renouncing and abandoning, albeit after its incidental discovery, an established nuclear weapons program. In terms of the vital nuclear fissile material element of this (and any other) weapons program, the availability to a proliferating agent of highly enriched uranium or weapons grade plutonium will not be conditioned by any step-change in the world's macro-security policy settings or change processes. In dealing, therefore, with the threats posed by the covert availability of weapons grade HEU and plutonium, solutions must be sought and found within the presently prevailing international security structures, architectures and institutions. The most perplexing conundrum, however, lies in understanding how to simultaneously and successfully enhance the status of the three aspects of nuclear fuel security noted earlier – transfer between programs, physical security and fuel cycle security – thus moving far nearer to closing the door on all aspiring proliferators before they manage to acquire fissile material.

We now move to consider the polar opposite of Libya's nuclear abandonment: the continuing evolution of Iran's nuclear energy development program, its origins, drivers and consequences, both for Iran and for the international community of nation states. The pathway to Iranian nuclear competency is complemented by its successful ballistic missile development program which together provide lessons for policy planners and others in understanding what impels states to seek the worst of all possible military solutions.

The Iranian Nuclear Program

The most worrying aspect of Iran's burgeoning nuclear development program, and one that goes to the heart of the response of many countries to its challenge, is encapsulated in the question "If not for the purpose of developing nuclear weapons, then *why* is Iran building a nuclear energy program with weapons development capacities?" Governments, security policy advisors and others with an interest in seeking a coherent explanation are led first towards consideration of reliably known facts and evolving intelligence about the nuclear program itself, as well as Iran's continuing ballistic missile programs. In this context, they are concerned (as with any potential nuclear proliferator) with Iran's scientific, technical and industrial potential.

With that aim in view, a Track 2 joint threat assessment, commissioned by the US-Russia Working Group on Counter-terrorism and Strategic Security (CTWG), has made a detailed and authoritative technical assessment of the Iranian nuclear program. Given the depths to which US–Russian Federation relations have recently fallen, especially over the now abandoned US proposals to establish a missile defense shield in Europe, this initiative has the ability to help repair the deficiencies of mutual trust still present in Washington and Moscow over US intentions towards its old adversary. It could also help to demonstrate to the governments of both countries that they *can* collaborate effectively on the most intractable strategic security issues of the day, including those addressing Iran's nuclear weapons program, as well as those of future aspirant state and non-state nuclear proliferators.[6]

The first point to make in discussing Iranian nuclear policy and action is that Iran is, as mentioned earlier, a foundation state party to the Nuclear Non-proliferation Treaty, having signed the NPT in 1968 and been subject to a Safeguards Agreement with the IAEA since 1974. As a Non-Nuclear Weapon State under the terms of the treaty, Iran has built a long-standing nuclear program on a foundation of pre-revolutionary assistance, principally from the US, France and Germany, through agreements for the construction of nuclear power plants and the provision of nuclear fuel. Ostensibly for peaceful purposes, that program is now more than fifty years old – a legacy of the US Atoms for Peace program of the Eisenhower era. Following its 1979 theocratic revolution, Iran received assistance from the A.Q. Khan network during the 1980s, and later from Russia in the form of a commitment to complete the construction of a major nuclear power generation facility at Bushehr. That facility had been abandoned by its German contractors in 1979 following the revolution, but was revived in 1994 with Russian help, and is expected to be in operation by 2010.

There is no doubt that Iran has, since 1980, been subject to the IAEA's inspection and verification regime against non-peaceful nuclear applications and materials transfer. From 2003, that regime included the more stringent, intrusive and unpredictable full-scope inspection requirements imposed on Iran following its voluntary accession to an Additional Protocol to its original Safeguards Agreements. Even so, Iran is now actively engaged in developing and operating a large-scale gas centrifuge-based uranium enrichment program, engendering deep and growing suspicion from its regional neighbors, the Western Alliance and globally that it intends to develop and deploy fission nuclear weapons using HEU or, possibly, weapons-grade plutonium-239. These suspicions are compounded by the fact that Iran is concurrently pursuing a comprehensive ballistic missile development and deployment program, together with, as yet, relatively primitive space launch capabilities (in February, 2009 Iran announced the orbiting of a small satellite into space).

---

6 Authoritative and contemporary information concerning the status of Iran's nuclear program can be found at: East-West Institute, *Iran's nuclear and missile potential: A joint threat assessment by US and Russian technical experts*, May, 2009, New York, NY. Available online at: http://www.ewi.info/groundbreaking-us-russia-joint-threat-assessment-iran-0. Accessed July 23, 2009. This paper is a joint threat assessment of the current Iranian nuclear threat scenario.

Iran is not a member state of the informal Missile Technology Control Regime (MTCR) which, through voluntary export control mechanisms, seeks non-proliferation of unmanned missile weapons delivery systems and technologies, including missiles with performance minima of 500 kg payload and 300 km range.[7] Nor is it a member of the more formal (but still legally non-binding) and complementary International (or "Hague") Code of Conduct Against Ballistic Missile Proliferation (HCOC), which seeks agreement among some 130 states on general ballistic missile policy principles, combined with modest commitments and limited confidence-building initiatives such as prior launch notifications.

Coinciding with Iran's strengthened IAEA Inspection and verification obligations, information provided during 2003 by Iranian political opposition groups suggested that Iran possessed nuclear-related sites which had not been disclosed to the IAEA. Subsequent inspections, no doubt assisted by Iran's recently expanded inspections obligations, revealed a covert Iranian program of undeclared nuclear activities covering the previous twenty years, and including uranium enrichment and extraction of plutonium from spent research reactor fuel. After a substantial delay, the first of four United Nations Security Council Resolutions, UNSC Resolution 1696 of 2006, demanded the immediate cessation of these activities, which by their commission and non-disclosure contravened Iran's obligations as a non-nuclear weapon member state of the NPT. It remains a matter of great concern worldwide that Iran has thus far steadfastly refused to do so.

Current Assessments

Iran is known to possess all of the major technical components necessary for the eventual completion of operational nuclear weapons through its nuclear energy production program. It has indigenous uranium resources estimated at some 900 tons, facilities at Isfahan to convert uranium oxide into uranium hexafluoride for centrifuge gas enrichment, substantial centrifuge cascade facilities at Natanz (with a target capacity of up to 54,000 centrifuge units), fuel manufacture and assembly facilities, also at Isfahan, and an almost complete and major energy production reactor facility at Bushehr (which is under full IAEA safeguards inspections) using Russian-supplied low-enriched uranium fuel.

Iran is also constructing a 40 MW "research" reactor at Arak which is moderated with heavy water and uses natural uranium fuel. This facility, due for completion by 2009, is similar to reactors operated by India and Israel for the production of weapons-grade plutonium, and it may be no coincidence that IAEA inspectors have been denied access to the facility since November, 2008.[8] Furthermore, the

7 Missile Technology Control Regime: *Introduction, 2009*. Available online at: http://www.mtcr.info/english/index.html. Accessed July 21, 2009.

8 World Nuclear Association, *Nuclear energy in Iran*, June, 2009. Available online at: http://world-nuclear.org/info/default.aspx?id=11522andterms=Iran+uranium+resource.

IAEA reported at that time that no progress had been made in negotiations with Iran on the possible military dimensions of its nuclear program.[9]

Although the watchword for all assessments of Iran's nuclear program and ambitions must be "uncertainty," the Track 2 report commissioned by the US-Russia Working Group on Counter-terrorism and Strategic Security (CTWG) noted above, and entitled "Iran's Nuclear and Missile Potential," made the following general assessment of the program:

- Iran has materials, equipment, technologies and expertise sufficient to produce low-enriched uranium fuel to approximately five percent for use in power generating reactor facilities. However, Iran has limited and possibly declining indigenous supplies of uranium, which may be a factor inhibiting Iran's plans for nuclear power generation into the future.
- Iran's increasingly capable uranium enrichment facilities make possible the production of uranium enriched in its fissile isotope U-235 to a weapons grade of ninety percent and above. Thus far, Iran is estimated to have produced, within one year, enough low-enriched uranium (LEU) to provide sufficient highly-enriched uranium (HEU) for one bomb. Future capacity will be determined by the number of centrifuge cascades which are eventually built, and the introduction of new, more efficient centrifuges.
- Evidence does not currently exist, nor is there any basis for *assuming*, that Iran has already accumulated weapons-grade fissile material, or that such material is being secretly stored. Similarly, there is no basis for assuming Iran's possession of LEU undeclared to the IAEA which is suitable for further enrichment, or in the amount required for obtaining significant quantities of uranium-235 (emphasis added).
- Since Iran has acquired nuclear technologies from Pakistan, but also from several European countries, in contravention of export control regulations, its need for further external assistance for its nuclear program cannot be discounted. This is an indicator of the need for strengthened controls over the export of dual-use technologies, including more efficient exchange of intelligence on such activities.

These assessments do not, of themselves, amount to an indisputable and current Iranian capacity to produce nuclear fission explosive devices. Far less does it amount to a capacity to produce a miniaturized nuclear weapon design which is sufficiently light and compact as to be deliverable by ballistic or cruise missile, or by attack aircraft. That capability would itself require a number of years of research and development to perfect.

As the CTWG assessment notes, Iran would, as a first step, certainly have to remove its production of weapons-grade fissile material from the eyes of the IAEA

Accessed July 21, 2009.

9 Supra. At Note 6, p. 7.

Inspectorate. Should Iran be able to do so, and also master other techniques unique to the manufacture of a fission weapon (such as producing HEU in centrifuges designed for LEU production, and converting uranium hexafluoride into uranium metal) it could conceivably produce an operable nuclear fission explosive device within one to three years. A further five years, and an explosive weapon testing program, would be needed to perfect, through miniaturization, operational weapons for delivery by ballistic or cruise missile.[10]

Iranian Missile Program Assessment

As noted above, Iran does not participate in either the MTCR or the HCOC informal missile non-proliferation initiatives. It does, however, possess a technologically competent, broadly directed, and growing program involving long, intermediate and short range missile development and deployment.

Iran's interest in ballistic missiles dates from the early nineteen eighties and its war with Iraq between 1980 and 1988, during which Iran is said to have fired more than 600 ballistic missiles at its adversary. Since that time, Iran has acquired ballistic and cruise missiles and their technologies from other, principally state sources, including Pakistan and North Korea. It has also developed them indigenously or in cooperation with other parties. It is important to note, however, that due to the high levels of security surrounding its missile activities, open source intelligence on the precise nature of Iran's missile development efforts is subject to considerable controversy.[11]

The United States has determined through intelligence estimates over many years that Iran may, at some stage, deploy ballistic missiles with intercontinental range, defined by the US as greater than 5,500 kms or 3,400 miles. There is no real consensus within the US intelligence community, however, about when that deployment might occur, since some contingencies – such as Iran's need for access to foreign assistance from North Korea or Pakistan – are essentially hidden from American view. So too is the status, capability and developmental potential of Iran's evolving space program and its operational space launch vehicle, the Safir. Nevertheless, growing threats such as Iran's undoubted prowess in ballistic missile development is a factor which loomed large in the US decision to deploy its Ballistic Missile Defense system, including the European component noted above.

To date, Iran has produced at least four liquid-fueled missile types with ranges varying from 315 kms to 1,100 kms, all using Nodong engines and other missile technologies supplied by North Korea. However, in order to be able to substantially extend the range of its future ballistic missiles, Iran will need to

10 Ibid. At pp. 4, 5.

11 Hildreth, S., Congressional Research Service, *Iran's ballistic missile programs: An overview*, February 4, 2009. Available at: http://opencrs.com/document/RS22758/2009-02-04. Accessed July 28, 2009.

introduce technologies which move well beyond their currently limited North Korean base. This is in contrast to Iranian claims that it can now field ballistic missiles with a range of up to 2,000 kms, sufficient to reach targets in the Middle East, Southern Russia and Southern Europe. While such a claim is probably an over-statement, one current estimate concludes that Iran now has the capacity to produce a missile with a 1,000 kg warhead and a range of 2,000 kms.[12]

The question of the Iranian space program presents further security conundra. On February 2, 2009, Iran launched a small satellite, named Omid, into low earth orbit, to the expressed concern of the United States and its European allies France and the United Kingdom. While this does not demonstrate a fundamental technical breakthrough towards an ICBM capability (since the first stage of the Safir launcher is still based on the North Korean Nodong missile) it does show that Iran's engineers may well possess the skills, knowledge and ingenuity necessary to design and produce multi-stage ballistic missiles, ultimately capable of carrying nuclear warheads. Omid's weight of 27 kilograms is well below the that of any operational nuclear weapon, but Iran's success in reaching earth orbit with its Safir technology is a significant step in laying the groundwork for indigenously developed intermediate (IRBM) and intercontinental (ICBM) ballistic missile programs.

Iran's growing missile capabilities constitute the second limb of broad international concern over Iran's nuclear intentions. Having achieved operational and miniaturized nuclear explosive devices, could Iran then use its missile force to pose a credible nuclear threat to its adversaries in the Middle East, Europe, and the United States? States in both regions, and the United States itself, can take comfort in the fact that, while it has shown a degree of competence in fielding what are essentially short range ballistic and cruise missiles, as well as a modest space launch booster, Iran lacks the depth of scientific, technical and industrial resources which has sustained missile development in the West, the former Soviet Union and elsewhere for many years.

Furthermore, Iran continues to be denied, through the Missile Technology Control Regime's export controls, and the international transfer interdiction work of Proliferation Security Initiative states, unfettered access to the wide range of highly specialized materials and equipment it needs to source externally in order to move into the first rank of missile producing states. Iran also lacks the necessary research, development and training institutions and infrastructure, as well as a sufficiently large, diverse and highly skilled workforce, all of which are essential prerequisites for Iran generating and sustaining successful indigenous missile programs in the IRBM and ICBM classes. Thus, Iran is currently limited to the performance of its Shahab 3M missile – with a payload of 1,000 kilograms and a range of around 1,100 kilometers – which nevertheless is notionally capable of delivering a single nuclear weapon.

---

12 Supra. At Note 6, p. 9.

In summary, the nature of Iran's nuclear energy and missile programs clearly show the path which a nuclear-aspirant non-nuclear weapon state party to the Nuclear Non-proliferation Treaty must follow if it is bent on presenting the world with a nuclear *fait accompli*. It may be that Iran has not itself determined to set out in that direction, but little of its nuclear activities in recent years affords legitimate grounds for optimism on that score. Iran's nuclear path is surely an extremely dangerous and unpredictable one, not least in view of the fact that when international relations between states enter the realm of the military application of nuclear technology, traditional certainties and decision making processes in government and beyond are inevitably subject to novel, untried pressures.

One outstanding example from history is provided by the resolution of the 1962 Cuban Missile Crisis, during which the United States came within a hair's breadth of all-out nuclear war with the Soviet Union. Speaking to President John F. Kennedy on the military option open to the US of attacking Russian IRBMs on Cuban soil, Secretary of Defense Robert McNamara told the president:

> I don't believe we have considered the consequences. I don't know quite what kind of a world we live in after we've struck Cuba, and we, we've started it .... how, how do we stop at that point?[13]

The strategic security implications for a state with nuclear policies of the kind exhibited by Iran, for its region and for the broader community of states, are grave and far reaching, at once both incalculable and immense. On one hand, its intransigence in the face of United Nations Security Council demands, and more generally those of the United States, the Russian Federation, and the states of Europe, is a direct challenge to the stability and future effectiveness of the Nuclear Non-proliferation Treaty. The NPT and its attendant galaxy of complementary international legal and political arrangements and activities – which comprise the nuclear non-proliferation regime – is increasingly threatened by states prepared to act as Iran has done. In breach of the most serious of international nuclear inspections obligations by refusing IAEA access to all of its nuclear sites under its Additional Protocol undertakings, Iran remains under partial surveillance by the IAEA Inspectorate, which still cannot verify that Iran is not undertaking a nuclear weapons development program.[14]

---

13 National Security Archive, *The Cuban missile crisis, 1962*, A National Security Archive Documents Reader, Chang, L. and Kornbluh, P., (eds), 1998 (rev. ed), New York, The New Press, 1998. Available online at: http://www.gwu.edu/~nsarchiv/nsa/cuba_mis_cri/declass.htm. Accessed July 31, 2009.

14 Implementation of the NPT Safeguards Agreement and relevant provisions of Security Council resolutions 1737 (2006), 1747 (2007), 1803 (2008) and 1835 (2008) in the Islamic Republic of Iran, *Report by the Director General, IAEA*, February 19, 2009. Available online at: http://www.iaea.org/Publications/Documents/Board/2009/gov2009-8.pdf. Accessed July 23, 2009.

A second dimension of the Iran nuclear issue is the potentially destabilizing effect in the volatile Middle East of the presence of a nuclear-armed Iran. A relatively sophisticated though flawed regional state in scientific, technical and industrial terms, Iran's future possession of nuclear weapons deliverable around the Middle East could without doubt push the governments of Saudi Arabia and Egypt, and perhaps others, towards nuclear weapons programs of their own. The effect of such a development on Israel's security posture is also of particular concern in the West, and specifically so with its American ally in view of the stated desire of the Iranian President, Mahmoud Ahmadinejad, to "see Israel eliminated from the pages of history".

A third, longer term, strategic security issue raised by Iran's pursuit of increasingly capable and long range ballistic missiles is the threat this will be perceived to pose beyond Iran's immediate region. States in the European theater, including the United Kingdom, were sufficiently concerned to welcome the US ballistic missile shield proposals for Poland and the Czech republic, despite serious Russian protests. The United States and the Russian Federation must also believe that there is, for they themselves, an equally significant security risk inherent in the burgeoning nuclear and missile technological capabilities of Iran. Should Iran be allowed to continue in an uninterrupted fashion, and finally achieve an operational nuclear strike capability, the United States, its allies and many other states may well view the period during which that outcome may have been prevented as an opportunity lost. Iran's success will not, of course, go unnoticed by other undeclared nuclear aspirant states around the world.

These three nuclear-relevant strategic security considerations are integral elements of short to medium term regional and global security assessments and planning. But what of the nature of current assessments, discussed earlier, on the future dimensions and potential of the Iranian nuclear program? Iran is acknowledged by the IAEA to possess extensive capacities in nuclear research and power engineering. It has operational uranium mines, it is continuing to produce low enriched uranium (LEU) using large scale centrifuge technology, it has three research reactors with a fourth heavy water reactor under construction and, finally, has conducted research into the use of weapons-grade plutonium (plutonium-239). It has also now produced around 1,000 kilograms of low enriched uranium from its centrifuge cascades at two enrichment sites.

Nevertheless, Iran steadfastly denies any involvement in preparations for nuclear weapon testing or production, despite extensive and very specific intelligence received by the IAEA from a range of member states suggesting that Iran is doing just that. That intelligence is believed to include research activities necessary to achieve an operable nuclear explosive device. Reported examples include studies of the behavior of high explosives (HE) used to initiate nuclear detonation; conversion of uranium dioxide into uranium tetrafluoride (which might indicate work on the preparation of uranium metal for use in a fission bomb); testing of high-voltage equipment for activation of HE detonators and Kryton switches for simultaneous detonation of shaped charges; development of guidelines for assembling and operating a detonation system; plans for the organization of underground tests; testing of a multipoint system for initiation of a high explosive unit of hemispheric shape for application in a lens implosion device; and biographical data showing the involvement of an Iranian expert in calculations of the radius of nuclear explosions.[15]

Strategic Consequences

In all of this, there are a number of assumptions and strategic realities which dictate the range of alternative approaches to the threat of a nuclear armed Iran. First, the United States must assume that Iran has, or will at some point acquire, the capacity and will to produce operational nuclear weapons deliverable on America's territory and that of its allies, including Israel and Western European states.

Second, the US needed to acknowledge (as it now appears to have done) that positioning a missile interceptor system in Eastern Europe was likely to have significant adverse strategic and deterrence consequences for its relationship with the Russian Federation, at least in the short to medium term. Furthermore, that the limited numbers of interceptors proposed for deployment (an initial figure of ten) would not be operationally effective against a determined attack using multiple delivery vehicles. To sustain a viable defensive posture would undoubtedly require a substantial increase in interception capacity and a move towards the deployment of more effective and politically acceptable mobile maritime missile defense systems.

Finally, it would be a mis-step for Iran's adversaries to assume, at this point, that Iran intends to undertake nuclear (or indeed conventional) warfare, either against the US, Russia or their respective allies, at least in the foreseeable future. As the CTWG's United States-Russia Joint Threat Assessment report, noted above, concludes, the most urgent task is not to rely on an inadequate and untried security system fielded against a possible nascent threat. Rather, it is for the United States and Russia to find the means by which Iran can convince the world that it is indeed embarked on nuclear research for exclusively peaceful purposes, and is not in pursuit of nuclear weapons. At some stage Iran must be allowed a definitive opportunity to allay the fears of the international community on this issue – an opportunity which is so clear that a failure by Iran to do so may be legitimately interpreted as an admission of the opposite aim.

In other words, Iran's future as a member of the international society of states will be secured, under the present circumstances, only through a comprehensive exercise in nuclear transparency. At present, it appears that Iran's leaders are satisfied that, on a calculation of costs versus benefits, their country's national interests are best served by pursuing a nuclear weapons option, rather than eschewing one. The

---

15 Supra. At Note 6, pp. 4–7.

nuclear "pull" factor has, in this case, outweighed the precautionary alternative approach.

Conclusion

The many detractors from the principles and operation of the 1970 Nuclear Nonproliferation Treaty often cite as the basis for their criticism the treaty's obvious and embedded distinction between those states which it recognizes as legitimate "Nuclear Weapon States," and those without that status. They often point out that the "bargain" which the NPT represents is in the nature of an unequal contract between two sets of parties with enormous disparities of global influence and power; the nuclear weapon "haves," in contrast to the "have nots". Only when all states enjoy real equality, verifiably and transparently guaranteed by international law, in their access to the benefits of applied nuclear energy, such critics believe, will the system (or society, or community) of states have a chance of achieving coordinated control over the misuse, for belligerent purposes, of the properties of fissile materials. India's often expressed position on its longstanding non-membership of the NPT is a strident example of this argument put into practice.

This fine aspiration ignores the realities of relations between states as they exist in the opening years of the new millennium. While the NPT has, it can be argued, played the key role in avoiding President Kennedy's specter of thirty or forty nuclear armed states around the world, its terms have never been altered in any way, nor are they likely to be. The world's state leaders, decision makers and their advisers must, then, act in respect of domestic and international nuclear policy issues in the context of present realities, which directly condition what is, in fact, achievable. They cannot afford, as the keepers of their state's national security, to take the risks attendant on any other approach. The case of Iran and its nuclear and missile programs, and the various international responses to them as discussed above, underscores this reality.

And yet, there can be little doubt that the best efforts of the United Nations, its nuclear agency the International Atomic Energy Agency, the United States, its friends and allies around the world, and many other states – whether nuclear or not – are progressively failing to a greater or lesser extent in their endeavors to rein in the insidious spread of nuclear weapons technologies, materials and expertise. As a result, there is a rising tide of states whose leadership is convinced, on balance, of the need to acquire these unbelievably powerful and inhumane devices. Those states have decided, or may soon decide, to follow the example of Iran rather than of Libya, and seek the doubtful benefits of nuclear weapons possession. Their choice will almost certainly be partially conditioned by the nature and success of measures directed towards Iran's, and North Korea's nuclear aspirations.

As noted in the Introduction to this chapter, the three most promising arenas in which global non-proliferation efforts should evolve are concentrated on the covert transfer of fissionable material from energy production to weapons programs; enhancing the physical security of all legitimately held and transported nuclear materials and technologies; and strengthening the security of nuclear fuels at both the front end (the enrichment of uranium) and back end (the reprocessing of, and extraction from, spent nuclear fuel of plutonium-239). Each aim involves complexly interlinked, contingent and even incompatible policy choices, but the broader targets are unmistakable: the denial of access to fissionable materials to those seeking to misuse them in the threat, or use, of nuclear explosives in all circumstances and at all times.

Chapter 7 takes the initial position that the world's nightmare scenario of active nuclear terrorism is a present danger and a growing menace. A specific item in the new Obama Administration's initiatives on homeland security and counter-terrorism is the strengthening of America's nuclear security through the enhancement of its detection technologies and architectures, as well as ensuring the security of US held nuclear material.

The chapter's approach is to clarify and carefully assess the suite of threats which attach to, and construct, plausible nuclear threats. It first reviews the evidence supporting claims in respect of the access of terrorist groups to nuclear materials and technologies, including enriched uranium and plutonium. This discussion is divided between access through the covert action of states or empowered individuals within them (A.Q. Khan of Pakistan being the prime exemplar) and access afforded through the use of false end-users, and other black market transfers, of dual-use technologies. Many of these have occurred with either the knowledge or assistance of legitimate Western corporations.

The chapter also reviews the nature and claimed scope of the terrorist threat created through access to nuclear technologies, contrasting this with the possible acquisition by terrorist groups of complete and operational nuclear devices. Finally, the chapter reviews the present range of responses to these perceived and actual threats – their efficacy and capacity to render nuclear terrorism difficult or impossible to accomplish.

Chapter 7
Nuclear Terrorism and Enriched Uranium

Introduction

The new Obama Administration in the United States has placed nuclear fuel security high on its priority international policy agenda. Breaking with the thrust of Bush administration policy, President Obama outlined a set of nuclear policy changes in a speech made in Prague, Czech Republic on April 5, 2009. In a wide-ranging public speech beyond America's shores, he outlined the renewed strength of American commitment to ending the shadow of nuclear weapons which continues to cast a pall across the world. In pursuit of a goal which, he admitted, would not be reached quickly or easily, he announced his administration's determination to work for a multilateral Fissile Materials Cut-off Treaty (FMCT) to verifiably end the worldwide production of fissile material intended for use in nuclear weapons. At the same time, he indicated new moves to strengthen the ability of the IAEA to perform comprehensive global inspections of nuclear-related sites.

Furthermore, President Obama declared his intention to press ahead with an International Fuel Bank of the kind discussed earlier, with the aim of reliably supplying the growing demand for nuclear energy while denying to states the capacity to indulge in weapons proliferation. In this latter aim, he stated that his administration will also seek the agreement of the international community to the punishment of states which violate their undertakings and legal obligations under the Nuclear Non-proliferation Treaty.

Finally, President Obama announced a new and ambitious international effort to secure all vulnerable nuclear material around the world within four years. Targeting the growing black market in nuclear technologies and materials, he urged the institutionalization of hitherto informal or non-binding initiatives such as the Proliferation Security Initiative (PSI) and the joint US/Russian inspired Global Initiative to Combat Nuclear Terrorism, launched in St Petersburg in 2006.[1] The new administration's broad counter-terrorist policy initiatives were, he intimated, closely linked with its nuclear fuel security policies. Among them, strengthening of the global non-proliferation regime, and promoting food security that fights world hunger and lifts people around the world out of poverty both clearly work, in the president's view, over the long term to help eliminate the worst

1 The White House, Office of the Press Secretary, *Remarks by President Obama, Hradcany Square, Prague, Czech Republic, April 5, 2009*. Available online at: http://www.whitehouse.gov/the_press_office/Remarks-By-President-Barack-Obama-In-Prague-As-Delivered/. Accessed August 8, 2009.

economic and social root causes of extremism leading to terrorist acts. Describing nuclear terrorism as "... the most immediate and extreme threat to global security," President Obama announced a Global Summit on Nuclear Security, now slated to be held in Washington in April, 2010.

As it clearly is for Obama, so the nightmare of nuclear terrorism is also the single most viscerally frightening large-scale threat present today in the minds of US national security planners, and equally so for many private citizens around the world. The heart and sum of all their horrors seem to be that, equipped as individual people are with varying levels of comprehension about the likelihood and probable nature of terrorist-initiated nuclear explosions, they believe that the governments of all states are essentially powerless to prevent a nuclear terrorist attack. Should the United States, or any other nation state anywhere in the world, become subject to this unthinkable act, there would, they surmise, be no chance of detection, no warning, no defense and no road to recovery from its effects.

In making assumptions of these kinds, they are unfortunately off the mark in many ways. Certainly, to take only one argument, the development and application of new and extremely powerful technologies in fields such as electronic intelligence gathering, digital information processing, remote sensing and surveillance, and the sheer bulk of experience now being garnered by counter-intelligence and counter-terrorist agencies world-wide have combined to successfully identify many nascent terrorist operations, and extinguish them.

This chapter is focused on the denial of fissile materials to aspirant nuclear terrorist groups. It does so by surveying the many ways in which possible nuclear terrorism is now being fought – from working towards upgraded and globally common physical security standards for nuclear weapons and weapon-grade fissile materials, to effective detection of illicit materials movements through border and transshipment controls, as required of all states under UN Security Council Resolution 1540.

Examples of failures by counter-terrorist agencies to detect and prevent terrorist operations are, of course, legion, and especially those which have been perpetrated in the West and resulted in the loss of many innocent lives. The bombings by al-Qaeda terrorists on Madrid's rail system in 2004 and the London Underground attacks of the following year fit this description. Those terrorist plots which are detected and closed down before an attack develops generally receive less extensive publicity but, in the context of building counter-terrorist intelligence and operational capacities for defeating subsequent attempts, are of equal or more valuable to law enforcement and other government agencies. One well known example is the alleged plan, revealed by counter-intelligence authorities in the United Kingdom in August, 2006, to detonate powerful liquid explosives smuggled on board up to ten trans-Atlantic passenger aircraft during flights between London and North America.[2] As anyone who regularly uses air transport is aware, the

2 Trans-Atlantic aircraft plot (2006), *The New York Times*, updated February 18, 2009. Available online at: http://topics.nytimes.com/top/reference/timestopics/subjects/t/

effect of this incident was an immediate and global tightening of restrictions on the carriage on aircraft of liquids and other materials whose containers are capable of use as a disguise for unconventional explosives.

Nevertheless, the operational planners of terrorist organizations are not in any way limited in the means, extent or planned effects of their actions while shaping new and unexpected attacks. Although they often tend to repeat earlier successful *modi operandi*, in many cases terrorist groups will attempt to circumvent existing counter-terrorist strategies by inventing innovative and ever more apocalyptic alternative strategems. Instead of a liquid explosives approach to the destruction of airliners, perhaps shoulder-fired guided missile might, they may believe, be more effective. Again, if a wide-scale attack on the mass transportation systems of major cities is likely, as such planners judge, to be of limited effectiveness, they may decide on a strategy based on the more certain and comprehensive mass destruction of people and national assets in heavily populated city areas. On that scale, the only possible effective solution may be judged to be to use the nuclear terrorist's path.

There should be no doubt in anyone's mind that committed terrorists are fully capable of making such a decision. There should also be no similar misapprehension that they would hesitate, for a moment, to use biological, chemical, radiological or nuclear means of achieving their aims if afforded the opportunity to do so. Of these alternatives, the nature of nuclear explosives – their overwhelming immediate and longer terms effects – makes the last of these option the most sought after of all.

This chapter lays out the most pressing of the challenges confronting counter-terrorist planners in responding to the nuclear terrorist threat in its various guises and international contexts. In doing so, it places a focus on the single most realistic and effective strategy available to counter-terrorist agencies – denial of access to the fissile materials without which a primitive nuclear explosive device is merely an exercise in machine-shop engineering.

Do Terrorists Have Access to Nuclear Weapons?

At the risk of reductionism, the short answer to this deceptively simple question must be "Yes, of course." In other words, there is a multitude of possible and feasible access points for terrorists bent on the pursuit of the ultimate explosive technology. This is as true in respect of the acquisition of complete and operational nuclear weapons and other explosive devices (ultimately, both fission and thermonuclear) as it is in terms of the entire suite of components, materials and technologies necessary to construct simple, and more complex, improvised nuclear fission explosive devices (INDs) from scratch – including obtaining highly enriched uranium or weapons-grade plutonium.

---

terrorism/2006_transatlantic_aircraft_plot/index.html. Accessed August 7, 2009.

In the first case, prevailing physical security deficiencies worldwide point to the nature of such access points, making what should be impossible – the theft of operational nuclear weapons – entirely within reach for fanatical groups instilled with determination and will, and equipped with the necessary means for its execution. While there have been great improvements, for example, in the security of Russia's nuclear weapon stockpiles and bomb-grade fissile materials over recent years, aided in large part by the United States, nuclear theft by both nuclear facility insiders, outsiders, or in cooperation, remains a continuing possibility. Among the factors acting to maintain unacceptable levels of risk of nuclear theft and subsequent terrorist acquisition from Russian sources are the presence within the Russian Federation of:

- widespread nuclear facility "insider" corruption and theft
- continuing risk of large-scale terrorist attack
- poorly trained and motivated conscript guard corps personnel
- under-investment in indigenously developed enhanced physical security measures
- weak security protocols and regulatory regimes
- a weak physical security culture[3]

These factors, especially concerning official corruption and the weakness of Russia's physical security culture, combine to present a clear and continuing danger that determined terrorists may yet acquire, and then detonate, operational nuclear weapons obtained from Russian sources. This is especially so since Russia is distinct from other major nuclear states in the globally unacceptable but resistant level of risk which its inadequate nuclear security performance still generates.

Nor is the United States itself immune from criticism leveled at its nuclear weapons security standards. A downward drift in morale within nuclear force units of the United States Air Force, for example, may have contributed to two separate and serious nuclear-related security breaches during 2006 and 2007. In the first incident, sensitive components of Minuteman III nuclear ballistic missiles were mistakenly transported from the United States to Taiwan before being recovered. In the second, "live" nuclear weapons were flown without appropriate authorization from Minot Air Force Base in North Dakota to Barksdale Air Force Base, Louisiana. Defense Secretary Robert Gates initiated a task force investigation into both incidents, which reported in October 2008 the worrying conclusion that the USAF was suffering "a serious erosion of senior-level attention, focus, expertise, mission readiness, resources, and discipline in the nuclear weapons mission."[4]

---

3 Bunn, Matthew. *Securing the bomb 2008*. Cambridge, Mass. And Washington, DC: Project on Managing the Atom, Belfer Center for Science and International Affairs, Harvard Kennedy School and Nuclear Threat Initiative, November 18, 2008. At p. 23.

4 Gertz, B., Air Force fails new nuclear reviews, *The Washington Times*, February 4, 2009. Available online at: http://www.washingtontimes.com/news/2009/feb/04/air-force-

In the second case above (the possibility of terrorist groups constructing crude fission explosive devices) and taking account of the wide range and variable performance of nuclear weapons security arrangements around the world, it is certainly not beyond imagination that a determined terrorist group could succeed in its quest. Once again, though, by far the most difficult element in such a project would almost certainly be the acquisition of highly enriched uranium or weapons-grade plutonium. Nevertheless, the example of the 2007 armed attack on the nuclear research facility of the South African Nuclear Energy Corporation at Pelindaba, situated west of Pretoria in the Republic of South Africa and discussed earlier, provides guidance. This attack amounts to cogent evidence of the capacity of a small group of armed attackers to penetrate the precincts of a supposedly impenetrable nuclear fortress. The facility, which houses the aging 20MW SAFARI "Atoms for Peace"-funded research reactor, is currently fueled by highly enriched uranium and used to produce medical isotopes. As a result, hundreds of kilograms of weapons grade uranium, a quantity also reported in the Washington Post as sufficient for some twenty-five uranium weapons, are legitimately stored at the facility. Pelindaba's attackers defeated several layers of security, including a 10,000 volt electric fence, to enter its control room and kill an operator with gunfire. This was a coordinated attack by technically sophisticated criminals, possibly with terrorist intent, that also included a separate armed attack by a second group which failed to gain entry into the facility from a different direction. Although the successful group did not penetrate the vault containing Pelindaba's highly enriched uranium stocks, the most important implication of this incident is clear: nuclear terrorism is far more than the fanciful notion of popular novelists.[5]

The Pelindaba event, and the still parlous state of physical security surrounding many of Russia's nuclear stockpile facilities, are active and passive examples of pathways available to aspirant nuclear terrorist organizations. A third, and more systemic, possibility is the agency of individuals and government affiliated agencies and commercial agents without provably direct links to terrorists, but which nevertheless have been causally instrumental in supporting their nuclear aspirations. This has often occurred through the removal of otherwise difficult or resistant barriers to the nuclear weapon option, not for terrorists themselves, but within nation states which have shown themselves prepared to make nuclear knowledge, expertise, technology and materials available to terrorist organizations.

Individuals and organizations of this kind have as their guide and inspiration the Pakistan-based nuclear smuggling group of the leading nuclear physicist A.Q. Khan, discussed in earlier chapters. They also include legitimate corporations based

---

fails-new-nuclear-reviews/print/. Accessed August 8, 2009.

5 Zenko, M., A nuclear site is breached: South African attack should sound alarms, *The Washington Post*, December 20, 2007, at p. A29. Available online at: http://www.washingtonpost.com/wp-dyn/content/article/2007/12/19/AR2007121901857.html. Accessed August 7, 2009.

in states of the Western Alliance such as Germany and the United Kingdom which have a history of exporting sensitive or restricted dual-use nuclear items by either ignoring their legal obligations or knowingly using false end-user documentation to avoid the detection of their export activities. Both phenomena, should they be allowed to continue and grow, will inevitably increase the number and value of the points of access available to terrorists seeking *all* of the elements of a nuclear explosives construction project.

The A.Q. Khan Case Revisited

The phenomenon of the nuclear smuggling activities of Pakistani physicist A.Q. Khan, who gradually built up a lucrative worldwide organization for the distribution of nuclear explosive technologies and materials to the highest bidder (and which, thankfully, was suppressed if not destroyed in 2003) has already been discussed, in general terms, in Chapter 4. Over some sixteen years, and involving in one way or another around twenty countries, A.Q. Khan's network sold Pakistan's nuclear technology – itself the product of Khan's effectiveness in obtaining it from commercial enterprises in established nuclear states such as France, Belgium and Germany – on a global scale black market. Khan and his network were only brought to book with the interception in 2003 of the German-registered ship *BBC China* and its cargo of uranium enrichment centrifuge components bound for Libya's nascent nuclear weapons program, as also noted in earlier chapters.

While the Khan network's activities no doubt did immense harm to the cause of global nuclear non-proliferation outcomes in terms of the effect of his activities on the nuclear aspirations of states, it is their secondary effects on the subsequent capacities and access of terrorist organizations in respect of nuclear technologies, materials and expertise which concerns us here. Khan has been characterized by former Director of the Central Intelligence Agency, George Tenet (many would say justifiably) as "... at least as dangerous as Osama bin Laden."[6]

The story which emerged into the public domain during 2003 and beyond was a scarcely credible chronicle marked by the demonstrated capacity of a single powerful individual, when ideally placed at the heart of his country's nuclear development efforts, to exert significant and durable agency and influence over international nuclear affairs. With the probable acquiescence or outright cooperation of the government of Pakistan, Khan did more during the twenty years preceding his arrest in Pakistan and the closure of his organization to harm the immediate prospects for global nuclear non-proliferation than any other agent, including the society of nation states.

---

6 Pakistan, A.Q. Khan, and the rise of proliferation networks, Introduction, *International Institute for Strategic Studies*, London, (at p. 7). Available on line at: http://www.iiss.org/publications/strategic-dossiers/nbm/. Accessed June 5, 2009.

In effect, Khan confirmed for those states harboring nuclear aspirations (including at various times Iraq, Iran, North Korea and Libya) the plausibility of their nuclear goals by making available all the necessary components for a "turn key" nuclear weapons program. For Khan's clientele, the prize of highly enriched uranium itself now presented no further insurmountable obstacles, in that conversion and gaseous centrifuge enrichment facilities were readily available, fully functional, from his organization. Having obtained the technical means to enrich their limited stocks of indigenous natural uranium, states such as North Korea and Pakistan itself had become outstanding potential sources of supply of HEU to any nuclear terrorist, or of the technical means to produce it themselves.

The success of the Khan organization is at least partially explicable in terms of the difficulties surrounding the control, by governments or international organizations, of the movement across international boundaries of nuclear related materials with dual-use properties. As discussed in earlier chapters, many nuclear items (and, of course, nuclear fuel itself) are equally necessary for use in the generation of electricity in production reactors *and* in the manufacture of nuclear explosive devices. As a result, a number of international governmental organizations have been established over time, such as the Nuclear (originally London) Suppliers Group, the Zangger (or NPT Exporters) Committee and the Wassenaar Arrangement for export controls over dual-use goods and technologies. The member states of these important global IGOs have cooperated to establish, through negotiation, export guidelines and trigger lists of restricted items, as well as undertaking the promotion of greater transparency and responsibility in the export around the world of dual-use items, those elements of a nuclear energy program which are also capable of contributing to illicit weapons development programs.

In a range of ways, then, and on a more positive note, the Khan experience has provided a benchmark for the future. Having been allowed to fester and evolve over many years, this highly organized and motivated nuclear smuggling network eventually became extremely adept at sophisticated strategies for the undetected movement of nuclear materials and components between nation states. In the process, and as its activities finally began to be unraveled and de-constructed after 2003, its demise revealed to national and international export control authorities and intelligence agencies the extent of their operational and methodological vulnerabilities to resourceful nuclear smugglers.

As the International Institute for Strategic Studies has recently found in a major study of the A.Q. Khan network, the intensive nuclear procurement activities of states such as Iraq and Iran, as well as the alternative and broader black market in nuclear related items which has assisted at least another nine nuclear aspirant states over half a century, opens up the field of complexity in the global nuclear market in a way which was far less clear in earlier times. This black market was in fact revealed as bifurcated into two major elements: the enormous and tightly coordinated Khan network, and the more heterogeneous, less structured component comprising a raft of smaller procurement networks, which had often been called into existence at the behest of states seeking invisible access to nuclear technology.

This second element nevertheless contained the full range of strategies available to the nuclear smuggler, including the suborning of international banking operations, the help of well paid European equipment manufacturing firms prepared to sign false end-user documents, the in-country advice and technical data supply of highly skilled nuclear engineers, and the like.

In the cases of Iraq and Iran, the Khan network established something analogous to a pair of symbiotic relationships with these two nuclear aspirants and adversaries by seeking to provide each with precisely what they individually sought. In return, Khan personally enriched himself. In the second instance, the phenomenon of the nuclear black market developed independently of the Khan network. Its clientele, as reported by the International Institute for Strategic Studies, may have included all or some of the following states: India, North Korea, Libya, Argentina, Brazil, Egypt, South Africa, Syria and Israel. Ironically enough, by unwittingly shining a spotlight of scrutiny on the full extent of global underground trade in nuclear related technologies and materials, Khan's influence has ultimately been, perhaps, as beneficial as it has been malign.[7]

With the putative disappearance of the Khan network, parts of which may yet re-emerge in new and unexpected guises, the world is left with a complex matrix of illicit trading activities in sensitive dual use nuclear items, as well as the possibility of the appearance of a significant black market in weapons-usable fissile material (primarily HEU and plutonium-239). The erstwhile Iraqi program attempting to develop nuclear weapons, which ceased in 1991, is a clear example of this phenomenon in operation in respect of a nation state, rather than a terrorist organization. However, in surveying the history of Iraq's clandestine nuclear program under the Saddam dictatorship, it is important not to lose sight of the opportunities which the global trade in nuclear weapons related technologies and materials presents to non-state organizations whose leaders and operatives are fanatically devoid of limits.

The Iraqi nuclear weapons quest ended in 1991 with Iraq's defeat in the First Gulf War, before which, however, Iraq had made intensive efforts to acquire technology and materials through a long list of foreign corporate suppliers (such as Matrix Churchill) which was later obtained by UNMOVIC, the UN agency monitoring Iraq's WMD disarmament from 1999. This list is, unsurprisingly, not available in the public domain. Following Israel's pre-emptive bombing of Iraq's Osirak reactor complex in 1981, the regime of Saddam Hussein had decided to forsake its vulnerable plutonium-based program for one using uranium enrichment, while remaining, through orchestrated subterfuge, ostensibly within the Nuclear Non-proliferation Treaty. In that quest, Iraq began to build a number of sophisticated and clandestine nuclear procurement networks. One authoritative estimate puts a

---

7 Ibid. See also Chapter 2, Nuclear black markets: Other countries and networks, p. 43 et seq.

cost of around US$10 billion on this nuclear weapons program, which eventually employed around 20,000 people.[8]

The nuclear program which subsequently developed involved a complicated web of inter-connected agents and functions; Iraq's diplomatic corps and intelligence agencies, overseas manufacturers, trading companies and middle men, foreign banks and Iraqi-owned front companies, all employed in an effort to defeat the dual-use export oversight of the international organizations mentioned above, and the (in the main) Western national security, intelligence and border control authorities.

There is insufficient space here for a close examination of the carefully modulated intelligence, diplomatic, organizational, transportational and financial techniques employed by Iraq in its search for the nuclear Holy Grail. However, the message which this wealth of intelligence on Iraq's nuclear program, as viewed from the inside and available in the public domain, sends in respect of the nuclear aspirations of well financed nation states such as Iraq and Libya is very clear. It is just as cautionary as a warning bell in considering the dreams of aspirant nuclear terrorists. If one looks long and hard enough, uses sufficiently well modulated and strategic subterfuge, and persists in the search, global black markets in any part of the world will eventually supply every necessary part of an operational nuclear explosive device. The terrorist's nuclear nirvana is entirely within reach.

Do Nuclear Terrorists Pose Credible Threats?

We must now go beyond general estimates and indirect analysis to look at how the various pathways of access for terrorists to nuclear explosive devices, while being feasible in principle, are in practice restricted and hedged around with numerous barriers to success – and as a result can be placed on a notional illicit nuclear hierarchy of feasibility from the readily practicable to the almost impossible.

The first point to reiterate and reinforce here is that it does indeed appear possible, on a reasonable and defensible basis, for a relatively well funded terrorist group with comparatively limited funding to assemble all the means necessary for the manufacture of an operational nuclear explosive device, or Improvised Nuclear Device (IND), *including* its fissile core. But can this really be done, on the ground and in reality?

In a well known 2006 article in the journal *Foreign Policy*, Peter Zimmerman and Jeffrey Lewis undertook an informed analysis in search of a coherent and pragmatic response to just this fundamental question.[9] They concluded that an Osama bin Laden-led terrorist group of some nineteen members could, with possession of sufficient enriched uranium and modest supplies of military items,

---

8 Ibid. At pp. 44 *et seq*.

9 Zimmerman, Peter G. and Lewis, Jeffrey G., The Bomb in the Backyard, *Foreign Policy*, November/December 2006.

most of which are readily available via the Internet or other open sources, employ a small team of committed terrorists with limited expertise to complete a crude but effective nuclear explosive device within a matter of months. Is this fanciful nonsense or the biggest ever wake-up call? Either way, the slightest possibility that such a feat may be even marginally feasible demands our closest attention.

As reported by Zimmerman and Lewis, in 1977 the US Congressional Office of Technology Assessment, which until its closure in 1995 provided members and committees of Congress with objective and authoritative analysis of the complex scientific and technical issues of the late twentieth century, estimated that a small group of relatively unsophisticated individuals, but including one with moderate technical capabilities, could build a workable nuclear fission device for a cost of around US$3 million at 1977 dollar values. Zimmerman and Lewis themselves estimated in 2006 that the total cost of a contemporary nuclear attack on America, using an IND constructed within the United States and including transportation and fissile material, need not exceed US$ 5.5 million.

Using standard and readily available machine tools (of the kind found in many machine shops around the world) they concluded that a small group of perhaps three or four individuals could fashion a "gun-type" device using a suitably refurbished military surplus recoilless artillery rifle. Such items are, they stated, readily available to those with the correct license within North America. The range of necessary skills within the full group of nineteen, consisting of a small team of physicists and a few equally small engineering teams (of perhaps three to four persons each), would extend to casting the uranium metal, modifying the artillery rifle, designing and building the device's structure, shaping the projectile and target elements of the supercritical uranium metal mass to be brought together in the bomb's "pit," overseeing the electronics package and finally conducting the explosion.

All this work would require the use of a remote rural property of perhaps a few hundred acres for a simple foundry, a basic machine shop and an electronics laboratory. Seclusion would also be needed to test the internal ballistics performance of the modified gun without attracting attention. At this point, Zimmerman and Lewis estimated, some seventeen people would have been fully occupied on the project for around one year.[10]

On these assumptions, which do not appear unreasonable or unobtainable in developed economies, the construction of a crude, though operational, nuclear fission device or IND seems, on the face of it, entirely feasible except, of course, for one omission: access to a sufficient quantity of appropriate fissile material. Osama bin Laden and al-Qaeda associated terrorist groups have long sought the nuclear option, and its operatives have made repeated attempts to procure fissile materials. As was extensively reported at the time, US military forces with embedded international news media personnel, in Kabul after the 2001 campaign to eject Taliban forces from Afghanistan, found extensive nuclear-related materials,

10 Ibid.

including crude fission bomb designs, which had been left behind in terrorist "safe houses" and other sites. There was no way of telling what materials and data they had taken with them.[11]

The conclusion of this analysis must be that the nuclear threat posed by terrorists is real, growing, and relatively resistant to countermeasures. The necessary materials, technologies and expertise, with the exception of suitable fissile material, are clearly available to any well-funded and determined person or organization which seeks them. This is, thankfully, not the case in regard to highly enriched uranium or weapons-grade plutonium, the denial of which is the single feasible and potentially most effective countermeasure which promises sustainable success, and that can be put in place over a period of no more than a small number of years. It is only in recent times, and now with the serendipitous and overt encouragement and commitment of the new American president, that the tide of nuclear fuel security enhancement has begun to turn.

Consider the extent of this change in nuclear fuel security policy direction in terms of the United States alone. President Barack Obama's Prague speech of April, 2009, reported above, committed his administration to work towards the long mooted but now moribund target of achieving a multilateral, eventually universal Fissile Materials Cut-off Treaty (FMCT) aimed at permanently ending the production of new fissile material intended for use in nuclear weapons. He also indicated greater US support for the IAEA's inspections and verification programs, as well as a four-year target deadline for securing all vulnerable nuclear material world wide, and the institutionalizing of currently informal international black market trafficking control initiatives, such as the Proliferation Security Initiative.

Furthermore, President Obama announced support for an International Fuel Bank to quarantine enriched uranium used for energy production away from potential proliferator states (and thus from some nuclear terrorists) as well as a move seeking international agreement to punish states which do not abide by their legal obligations under the Nuclear Non-proliferation Treaty. Significantly, he also announced a Washington Summit Meeting on Nuclear Security, now to be held in April, 2010.

Meanwhile, other international initiatives supported by the United States, will remain active under the Obama Administration, such as the US–Russia-based Global Initiative to Combat Nuclear Terrorism, announced in 2006, and the earlier and complementary G8-based Global Partnership Against the Spread of Weapons and Materials of Mass Destruction, announced at the G8 Meeting held in Kananaskis, Canada, in 2002.

---

11 Albright, D., Al Qaeda's nuclear program: Through the window of seized documents, *The Nautilus Institute Policy Forum Online, Special Forum 47*, November 6, 2002. Available online at: http://www.nautilus.org/archives/fora/Special-Policy-Forum/47_Albright.html. Accessed August 9, 2009.

Nuclear Counter-terrorism: The Policy Task

While the new Obama presidency promises a new approach to nuclear fuel security policy for the homeland security of the United States against terrorist attack, and hence on a world scale, it is vital to understand that the fundamental components of the nuclear threat which terrorists pose to the world have been relatively slow to evolve. As a result, the basic premises which condition the set of possible countermeasures available to combat them are also reasonably clear, having resolved only slowly into sharp relief. In the case of nuclear fuel security, restrictions inhibiting terrorist acquisition are likely to be technical, financial and political in nature. In a field which demands coordinated and complementary international action if it is to be effective, as always it is the broad political acceptability of coordinated strategems which often delay or deny their deployment.

An authoritative starting point for considering the effectiveness of contemporary and future nuclear counter-terrorism strategies is to test the potential efficacy of a new strategy for reducing nuclear threats, including terrorist-based threats, against a set of renewed US policy directions. As the new Obama Administration takes up the reins of executive power, those enunciated in the 2007 report of the National Security Advisory Group of the Belfer Center for Science and International Affairs, John F. Kennedy School of Government, Harvard University, entitled "Reducing nuclear threats and preventing nuclear terrorism," provide a substantial foundation, since the Group's report is the work of a number of senior scholars, as well as former senior US government and other political and diplomatic figures, all experienced practitioners or experts in nuclear security issues.

In pursuit of the overarching aim of protecting the United States, its allies and friends from nuclear attack, and from coercive pressures by states possessing nuclear weapons, their report recommends the following goals for such a strategy:

- "As our highest priority, prevent terrorists from acquiring and using nuclear weapons against the US homeland or against US friends or forces abroad
- Roll back North Korea's nuclear weapons capability and head off Iranian efforts to acquire nuclear weapons;
- Discourage and prevent additional countries, even US friends and others who might be tempted by recent setbacks in the nonproliferation effort, from acquiring nuclear weapons capabilities;
- Maintain effective military forces, both nuclear and conventional, capable of deterring countries that already possess nuclear (or biological/chemical) weapons from using such weapons against the US, its allies, and friends or from using those capabilities to pressure or undermine US friends and interests around the world;
- Reduce the likelihood that nuclear weapons will be used – whether by Russia, the US, or any other nuclear power – as a result of accident or misperception;

- Ensure that any large-scale expansion of civil nuclear power programs worldwide will proceed within the framework of new rules and procedures that can minimize the risks of proliferation to additional states and terrorists".[12]

These six goals are thus modulated in such a way as to encompass nuclear threats emanating from both nation state and terrorist sources. The most difficult to attain, at least in principle, may prove to be that accorded the highest priority of all – the prevention of terrorist from acquiring and using nuclear explosives against the American homeland, her allies and overseas forces.

In pursuit of the six targets, the Group's members urge consideration of the following ten broad policy focuses:

- Securing nuclear materials world wide
- Protecting the homeland against nuclear threats
- Stopping North Korea and Iran
- Ensuring credible deterrence
- Modernizing US deterrence forces and infrastructure
- Revising US declaratory policy
- Pursuing missile defense
- Reinvigorating arms control
- Shoring up the non-proliferation regime
- Promoting the safe expansion of civil nuclear energy.[13]

Although some may question the need, it is helpful to recognize that this agenda for action is clearly proposed within a world view which accepts the necessity of tackling both state-based nuclear weapons proliferation *and* the nuclear terrorist threat as a single challenge. Rather than approach the global nuclear security policy task from the perspective of simultaneity, or of the complexity of two related but separate challenges, it proposes to deal with *all* manifestations of nuclear threat together. This does not demand policy homogeneity, but rather the capacity to imagine and build in flexibility and integrated policy solutions to complex issues which inter-mesh both state and non-state nuclear implications and consequences.

The list of strategic goals for the re-emergence of a world safe from the shadow of nuclear weapons (and incorporating, in terms of this part of the study, nuclear terrorism) together with the ten policy targets proposed in response must be assessed in two ways. Firstly, by qualitatively evaluating the probable effectiveness of each element in both categories (aims and methods), and secondly

---

12 Carter, Ashton B., et al., National Security Advisory Group (NSAG). *Reducing Nuclear threats and preventing nuclear terrorism.* October 2007. Available online at: http://belfercenter.ksg.harvard.edu/publication/17635/reducing_nuclear_threats_and_preventing_nuclear_terrorism.html. Accessed August 11, 2009.

13 Ibid.

by generating hierarchies of urgency in order to evaluate the most important opportunities for policy synergies which bring added efficiency to the twin goals of denying opportunities for both state-based nuclear proliferation *and* nuclear terrorism as a single policy target.

The final stage of the analysis comprises judgments about the success, or failure, of existing responses, especially as they impinge on the nuclear terrorism threat. Examples include a wide range of initiatives from the US itself, as well as the US in conjunction with allies and other states; from agencies of the United Nations and other elements of international civil society (such as the many non-governmental, privately funded and specialist policy advocacy organizations with global presence); from new and innovative international law and norms, including customary law; from the emergence of new regimes governing the limits of acceptable international behavior; and from the adaptation and enlistment of new technologies of all kinds, with examples ranging from high speed data processing to remote surveillance and detection techniques, and new, modular and proliferation-resistant production reactor designs.

These initiatives immediately generate myriad questions, many as yet unanswerable. Do these and other methodologies adequately respond to the unthinkable threat of a nuclear catastrophe? Are others required in order for the society of states to be as certain as reasonably possible that the nuclear threat has been reduced as far as it can be? How far has the world progressed in cooperating towards nuclear non-proliferation generally, including suppressing nuclear terrorism, and what evidence is their in support of claims of progress? Answers to questions of this kind will go some way to showing what still lies unaccomplished in the race to deny the unimaginable to enemies without limits (some of the most important prospective nuclear fuel-related policy options will be reviewed in Chapter 8).

Evaluating Nuclear Counter-terrorism Strategies

The first question for resolution is whether the six broad goals of a US preventative nuclear security strategy outlined above fit the bill in terms of a nuclear fuel and vulnerable weapon-based counter-terrorist methodology.

The first and highest priority goal is rightly that of preventing terrorists from acquiring and using nuclear weapons against the US homeland, or against US friends and forces abroad. It is also the single most important national security focus, as identified by the United States Department of Homeland Security in its Strategic Plan, 2008-13, in terms of preventing terrorists from entering the United States.

The rolling back of the nuclear weapon programs of North Korea and Iran, together with success in inhibiting new entrants in the nuclear weapons proliferation stakes, have met with limited success as strategic policy goals. They are also, for the new Administration as for its predecessor, policy "givens" which, though vital, do not progress policy innovation when considering the flexibility and small scale dynamics of opportunities for nuclear outrages when presented to terrorist groups. A continuing and traditional nuclear and conventional US deterrence policy and posture, as outlined in the fourth strategic goal above, is firmly in the same category, as is the essential need to guard against the dangers of misperception in a US-Russia nuclear "launch on warning" confrontation scenario.

The final strategic nuclear security policy goal above has, however, identified the core issue for a strategy which seeks to "[e]nsure that any large-scale expansion of civil nuclear power programs *worldwide* will proceed within the framework of new rules and procedures that can minimize the risks of proliferation to additional states and terrorists" (emphasis added).

There is little doubt in 2010 that a renaissance is under way in the use of base load nuclear power generation, in the face of increasingly strong evidence pointing to global atmospheric warming and dangerous climate change phenomena. As a reliable, low emission and economically feasible substitute for commercial scale energy production, generated by burning hydro-carbon laden fossil fuels, nuclear power is now experiencing a strong renewal of interest and significant new capital investment. This is so from China through Western Europe, on to the United States and well beyond.

Given the burgeoning global nuclear power industry, there is an unequivocal and urgent need for new generating technologies which make access by terrorists to fissile materials all but impossible. To re-engineer and modify the commercial nuclear fuel cycle, on a world wide basis, will be, in itself, technically and diplomatically difficult in the extreme. Nevertheless, it must be accomplished if the economic promise of nuclear power is ever to be fully realized in a reliably secure way.

What this brief analysis of the six mooted strategic goals above does, above all else, though, is to point up the dangers of failing to renovate well-worn, traditional and proven policy pathways while also not fully re-imaging the extent of new, related but very distinct and difficult nuclear challenges. Of course the United States must continue to keep its citizens safe. This is always the first and over-riding duty of its leaders and governments, as it is of the political leadership of every nation state. Of course it follows that the US must maintain its nuclear deterrent posture, in the knowledge that only leaders of nations wishing to commit national suicide on their people's behalf will attack the US homeland with nuclear weapons.

Unfortunately, this is precisely where the most dangerous threat to the American population most assuredly lies; not from national leaderships (so far as is known) but from the leaders of international terrorist organizations, in their determination to attack the heartland of their sworn enemy without any necessary assistance from the government or resources of a nation state. This is the point at which US nuclear security policy and, especially in counter-terrorist terms the nuclear fuel security policy embedded within it, must be re-set.

No longer is it either prudent or possible to exclude from an integrated policy amalgam of this kind the possibility, however remote it may seem, that nuclear terrorism could strike the American heartland. Instead, all nuclear security policy settings must be conditioned, and take account of, the need to ensure that bomb-suitable fissile material, including highly radioactive fission products found in stored reactor spent fuels, is not clandestinely obtainable by any agent whatsoever. Although the detonation by terrorists of a "dirty bomb" using fission products such as Cesium-137 has not been canvassed in this discussion, the distribution of such highly toxic material using conventional explosives in densely populated areas must not be discounted.

As to the ten mooted nuclear security policy focused areas above, it is difficult to criticize a list such as this in the potential or importance of each entry in an integrated response to state and terrorist-based nuclear threats. However, a more useful approach may have been to take the most overwhelmingly urgent of *all* nuclear threat – the possibility of a nuclear attack upon the US, its people and territory – and proceed on the basis of the capacity of responses to *that* threat in forming the basis for resolving more traditional and better understood nuclear threats emanating from state nuclear proliferators. As this analysis has consistently maintained, nuclear security is not ideally comprehensible in its form and resolution by analysis which merely divides threats according to their authors. Given this alternative approach, what would a revised list of US priority policy areas look like? The following list is intended to convey a flavor of the theoretical complexity of the policy challenge, as well as the difficulty of its simultaneous or even consecutive resolution:

- Make fully secure all nuclear weapon stockpiles world wide, as required by United Nations Security Council Resolution 1540
- Eliminate all unnecessary stockpiles of nuclear materials world wide, including all unnecessary highly enriched uranium and weapons grade plutonium stocks, and consolidate those remaining
- Conclude a Fissile Materials Cut-off Treaty preventing the production of new weapons grade fissile material
- Ensure, through UN-based sanctions and other punishments, that all states party to the Nuclear Non-proliferation Treaty abide by their legal obligations under the terms of the treaty
- Provide the International Atomic Energy Agency with the resources necessary to permit it to verify through inspection processes the non-transfer of all nuclear related technologies and materials from civilian to military nuclear programs
- Develop universally enforceable and transparent physical security norms governing the movement of all fissile and other nuclear related materials and equipment within states' territories, during maritime transportation and across international borders

- Work towards the interdiction of all illicit transfers of nuclear materials, technologies, data and expertise
- Work towards a universal Nuclear Weapons Convention banning all nuclear weapons in perpetuity, and ratify the 1996 Comprehensive Test Ban Treaty
- Eliminate all separation of weapons usable civilian plutonium from spent fuel sources
- Promote the universal introduction of new, proliferation resistant reactor technologies in which plutonium separation is not possible
- Institute an International Fuel Bank or Banks, to quarantine enriched uranium for power generation from the possibility of its illicit transfer to weapon programs
- Make fully secure all holdings of HEU and plutonium within the United States, including research, medical isotope and weapons materials.

It would be naïve in the extreme to suggest that a policy agenda of this kind, shorn as it is of references to the traditional construction of state-based nuclear threats, is complete. But perhaps it is a suitable starting point for consideration of a much shorter, but still traditional list than aging Cold Warriors may be used to, in terms of any residual or tenacious nuclear threats which have no practical connection to nuclear terrorism.

However, it only remains in this chapter to note the moderate but welcome levels of success enjoyed by the range of US policies currently deployed or contemplated. In the nature of policy evolution, the story is, of course, a mixed one. Some claimed successes (for instance, the work of forces operating within the framework of the Proliferation Security Initiative), inhabit the same policy space as examples of promises unfulfilled (in the case of the stalled proposal for a twenty year US-Russian moratorium on plutonium separation), and significant policy failures (including failure to rein in the continued pursuit of nuclear weapons by Iran, and renewed defiance by North Korea, a known disseminator of nuclear related materials, in the form of a defiant second nuclear weapon test).

In each case, and in many other examples of nuclear non-proliferation and nuclear counter-terrorism work by the US and around the world, there is a perceptible move in emphasis towards the paramount need to frame future non-proliferation regime policy in the direction of nuclear fuel security. Although only one part of the nuclear proliferation story, the wide distribution of weapons-grade fissile material, and the technologies with which to produce it (especially HEU) mandates its unique importance. Take away its acquisition and the mere possibility of a nuclear detonation disappears.

It is to be hoped that the log-jam of nuclear initiatives, many of which have been held in abeyance during the era of the second Bush Administration, can now be cleared away by President Obama's elevation to the presidency. Issues such as US ratification of the Comprehensive Test Ban Treaty, a necessary step before the treaty, now with over one hundred and eighty signatory states parties, can enter into force cannot usefully be postponed any longer in America's national security interests. President Obama has already indicated that the United States will now move towards that goal.

On an even more positive note, the single most potentially crucial new policy initiative, and one which dates from the Bush era, is the proposal (one among a dozen such proposals from around the world) to establish a universal International Nuclear Fuel Bank, as noted earlier. This single proposal, if successful (and rapid acceptance cannot be reasonably anticipated) should do more than any other policy initiative can to finally eliminate the dark presence of nuclear terrorism from the earth. This and other competing Fuel Bank proposals will be canvassed, contrasted and reviewed in their potential to deny what the nuclear proliferators and terrorists so crave – the unimaginable terror available to them from the malign use of nuclear fission.

Chapter 8 continues to review the future of nuclear fuel security and its evolution by placing the issue of international control over the enrichment of uranium, and the availability of weapons-usable plutonium, in this broader context.

Chapter 8
Old and New Strategies for Non-proliferation Progress

Introduction

This chapter comprises a review of current and progressing strategies for global nuclear non-proliferation, with stress on those which use nuclear fuel security as the fulcrum around which future comprehensive non-proliferation success will pivot. It places the issue of independent international control over the enrichment of uranium for power generation, and the availability of weapons-grade plutonium, within this broader context. Furthermore, the global nuclear fuel bank concept canvassed in the previous chapter will be developed and evaluated in its potential for non-proliferation effectiveness. The chapter uses an integrated approach towards this and other non-proliferation efforts in the belief that all possible and viable options must be canvassed, tested and – if found to be effective – applied in the cause of countering the unchecked spread of nuclear weapons technologies.

As the titles of this book suggests, the existence of ineffectively controlled and secured enriched uranium and weapons-grade plutonium has the capacity, in and of itself, to exert an inordinate degree of influence over the security status of America's homeland in the coming years, as indeed it has done now for decades. While this certainly remains true, the pathways of leverage for those individuals or non-state groups who should never be allowed to exert effective control over fissile material are very different from those of earlier times.

As the rising threat levels associated with inter-linked nuclear weapons proliferation and nuclear terrorism become ever more serious, so cooperative and collaborative international policy solutions, which, during most of the nuclear era have been universal anathema, are now being revisited with ever more serious intent. So serious, in fact, that nuclear policy advisers within and outside government, both in the United States and around the world, as well as within privately funded policy research and advocacy organizations (with the "think tank" appellation coined by defense analyst and Hudson Institute founder Hermann Kahn) have in recent years begun to canvass the previously unimaginable.

The concept of international fuel banks to quarantine the enrichment of uranium and, eventually, the separation of plutonium from spent reactor fuel is an idea whose time, most definitely, has now come. When once translated into practice with the universal acceptance of states, the reinforcement of international treaty law, and the transparency of independently verifiable international control over the global disposition of fissile material, they represent the single most important tool for practical and sustainable progress in reaching the ultimate goals of global nuclear non-proliferation. It is important to understand, though, that such an initiative should in no way be regarded as a substitute for the wide range of nuclear fuel security measures already canvassed, such as a universal and comprehensive Fissile Material Cut-off Treaty, or the down-blending of HEU made redundant by the increasing rate of conversion to LEU of research reactors in many countries. Rather, if successfully introduced, an international fuel bank (or banks) has the potential eventual to consolidate and reinforce much of the necessary but fragmented international effort now devoted to bringing fuel security levels to minimally acceptable international standards. In the interim, much still remains to be accomplished in bringing international fissile material security to levels which are then sustainable with both the effective deployment of existing measures (many of them currently in the form of US or joint US-Russian initiatives) and the immense organic contribution to the nuclear non-proliferation regime which an international fuel bank architecture and operational framework would bestow. What, then, does the concept of international nuclear fuel banks amount to?

The International Fuel Bank Concept

In the opening years of the nuclear era, as memories of total war translated into the challenges of the peace, the optimism and even idealism which accompanied the advent of the United Nations Organization rapidly faded. The promise of a globally active and supreme expression of the will of the nation states, guaranteeing that there would be no return to total war in an era of nuclear weapons was, as now seems inevitable with the benefit of hindsight, rapidly replaced by the real threat of just such a tragedy.

Even so, the struggle for international control over nuclear weapons technologies began with the swift and total failure of the United National Atomic Energy Commission, instituted early in 1946 by the first resolution of the UN General Assembly, to make meaningful headway over international control of the atom. As the only possessor of nuclear weapons at that time, the United States declared its willingness to destroy them only on condition that international control over nuclear developments would not be subject to a Security Council veto. Confronted by the hardening and mutual suspicion of the United States and the Soviet Union over the possession and potential use by each of nuclear weapons, the UNAEC's experience showed how difficult the road ahead would be for nuclear disarmament and non-proliferation by adjourning indefinitely after three fruitless years.

Although the International Atomic Energy Agency eventually emerged a full decade later, it did not, of necessity, aspire to the idealistic heights visited by its predecessor, instead reflecting a more circumspect exercise in the art of the politically possible. Its success, based largely on its acknowledged expertise, functional autonomy and near universal support of the nation states, remains unchallenged in the early years of the twenty-first century. For these reasons the IAEA presents the world with its most promising opportunity to use the only nuclear-focused organization with global nuclear reach as the catalyst, agent and guarantor for a controversial but vital long term response to the nuclear fuel security conundrum: the concept of the internationally controlled nuclear fuel bank. While the success of this concept will be neither rapidly nor easily attained, it is far too important for the world's future, in which nuclear power will play an increasingly important role, not to be pursued with every resource available to the society of states.

The recent history of the fuel bank concept has followed a familiar path as states began, from a slow beginning, increasingly to understand the potential benefits available from controlled access to nuclear power.[1] From 2003, when (then) Director General of the IAEA, Dr Mohamed ElBaradei, called for urgent discussion among member states about how best to move forward on the issue, some twelve proposals were put forward. Many variations on the theme of the internationally-based supply and control of LEU were advanced, from establishing an IAEA-controlled "last resort" reserve of low- enriched uranium, to providing alternative forms of backup assurance of supply and setting up international uranium enrichment centers.[2]

In mid-2007, for example, the Russian Federation proposed an "International Uranium Enrichment Center," to be established at its Angarsk Chemical Complex in Siberia (already an LEU production facility), be operated under IAEA safeguards, and be dedicated to the low enrichment of uranium for international power reactor fuel use. This Russian nuclear fuel cycle service would be guaranteed to any state in full compliance with its obligations under the Nuclear Non-proliferation Treaty, although the material itself would stay under full Russian national control while on its territory.

A year later, Germany, the Netherlands and the United Kingdom, having held their important Berlin Conference entitled *Nuclear Fuel Supply: Challenges and Opportunities*, proposed to the IAEA a similar scheme to "multilateralize" the fuel cycle by developing guarantee arrangements for the international enrichment of uranium. Although these three collaborating states did not see their proposal as embodying any element of coercion, nor presenting a challenge to the rights of states under the NPT, they did express a clear view on the four key requirements for a successful nuclear fuel bank:

1 *Multinational fuel bank proposal reaches key milestone*, International Atomic Energy Agency, News Centre, March 6, 2009. Available online at: http://www.iaea.org/NewsCenter/News/2009/fbankmilestone.html. Accessed August 15, 2009.

2 *Nuclear energy, the need for a new framework*, Statement of the Director General, International Atomic Energy Agency, Berlin, April 17, 2008. Available online at: http://www.iaea.org/NewsCenter/Statements/2008/ebsp2008n004.html. Accessed August 15, 2009.

- multinational control
- access to fissile material on a non-discriminatory basis
- release of fissile material using criteria that are not politically motivated
- an overarching multilateral framework linking non-proliferation and nuclear disarmament

Germany, the Netherlands and the UK noted that their proposals were likely to be opposed in favor of building national fuel cycles, a preference which would be justified on ground of self-sufficiency, the global status available in terms of political, economical and technological independence, and the need felt by a number of national governments to master the nuclear fuel cycle in the context of the right of states to do so under Article IV of the NPT.

On the obverse of the coin, the three proponent states pointed out the disadvantages flowing from opposition to their proposed fuel security measures, namely increased proliferation risks, the high cost of establishing national uranium enrichment facilities, and the fact that the commercial supply of enriched uranium is likely to remain both adequate and affordable into the foreseeable future. They also noted the crucial role played by trust, which must necessarily be invested in the supplier state or states by clientele states (despite the proposed presence of IAEA and other international oversight arrangements).[3]

Director General ElBaradei, in a speech to the 2008 Berlin conference, laid out his own vision of the process by which the success of any nuclear fuel bank proposal could be expected. *First*, a system for the assurance of supply of suitable reactor fuel, and if necessary the reactors themselves, must be established. *Second*, all new enrichment and reprocessing facilities and activities must be placed under exclusively multilateral control. The *third* step, Dr ElBaradei emphasized, would be to place all existing national enrichment and reprocessing facilities under multilateral control. An absolute prerequisite for all three stages of this process, he asserted, would be the conclusion of a universal Fissile Material Cut-off Treaty (FMTC) prohibiting all future production of weapons grade material.

The IAEA Director General concluded his remarks with reference to the single most promising of all the fuel bank proposals received – that developed by the private sector Nuclear Threat Initiative (NTI), a security policy advocacy and charitable organization funded initially by private individuals, and more recently by the United States and many others states. An outstanding example of private sector initiative and philanthropy, and one perhaps without peer in its significance for global security in any era, the NTI International Fuel Bank initiative was announced in September, 2006. A generous initial NTI contribution of US$50 million, funded by the philanthropist Warren Buffet, was made conditional on the receipt by NTI of additional funding of US$100 million, or low enriched uranium to that value, before September 30, 2009. That condition was fulfilled

3 IAEA INFCIRC 728, June 12, 2008. Available online at: http://www.iaea.org/Publications/Documents/Infcircs/2008/infcirc728.pdf. Accessed August 15, 2009.

in March, 2009 with a US$10 million contribution from the government of Kuwait, bringing total initial funding for the nascent IAEA-led NTI fuel bank to US$157 million, including contributions from a total of thirty-one states. The bulk of state support was provided by the United States (almost US$50 million) and the European Union (a further US$33 million).[4] Other significant contributions included those of the United Arab Emirates ($US10 million) and Norway (US$5 million).

The NTI nuclear fuel bank proposal is closely aligned with the vision expressed in 2008 by former IAEA Director General Mohamed ElBaradei, and noted above. This includes the crucial elements of a guarantee, based on IAEA control, of non-discriminatory, non-political, last resort supply of fuel to states that have been denied all other sources of fuel, but are nevertheless in full compliance with their NPT obligations. It is necessary here to reiterate that there is an inevitable and pragmatic gap between the most desirable final position, in which all nuclear fuel supply operations and services are eventually brought under the permanent aegis of the IAEA in one way or another, and a situation in which IAEA-controlled fuel supply arrangements are initially in place only as an impartial supply of last resort. Such an initial arrangement would amount only to a dependable "safety net" for states with alternative energy source vulnerabilities and temporary or unsustainable fissile material supply difficulties. The eventual transition of a limited IAEA-led global nuclear fuel bank institution of this kind into one with real and unchallenged capacities to act as an exclusive and universal nuclear fuel bank will require trust and political will on a herculean scale, but it must be attempted nevertheless.

Only by implementing an incremental and carefully modulated approach with this aim in view is it likely that the ultimate goal of such an international nuclear fuel bank, or network of fuel banks, will be reached. Until then, there is an urgent need to build sustainably high levels of trust in a revolutionary concept which has deep and lasting implications in the minds of many, including state leaders, for state sovereignty, political independence and international prestige. When considered in this light, the question of mutual trust, to be exercised by nation states in the field of international control over nuclear fuel, seems fraught with political difficulties, if not practically impossible to reach. Even so, without an operational multilateral nuclear fuel enrichment architecture in place, it is doubtful whether the necessary levels of trust in the reliability of the mechanism itself, much less the motives of those states involved in its origins and evolution, will be attainable for states which genuinely do not have any interest in the military use of the fissile atom.

And yet, there is no functional reason to dismiss the fuel bank concept as unworkable *prima facie*. All states have the example of the 1970 Nuclear Non-proliferation Treaty before them as a model and template for the success of

4 Press Release, NTI/IAEA Fuel Bank Hits $100 million Milestone; Kuwaiti Contribution Fulfills Buffet Monetary Condition, Nuclear Threat Initiative, March 5, 2009. Available online at: http://www.nti.org/c_press/release_Kuwait_Fuel_Bank_030509.pdf. Accessed August 15, 2009.

multilateral nuclear treaties borne of urgent need. Certainly, as an artifact of human endeavor, the NPT is imperfect in its intent, structure, interpretation over time, review and application, and has been so now for four decades. Nevertheless, its strength has certainly been demonstrated on many occasions through the flexibilities which reside in its interpretation, rather than its 'black letter' international law character. Over nearly forty years, the NPT has been transformed from a nascent and ambitious United Nations-led multilateral nuclear treaty, the result of urgent negotiations at the Eighteen Nations Disarmament Committee in Geneva between the United States and the Soviet Union, into a global agreement which lies at the center of a complex regime of international nuclear non-proliferation activities, whether legal, technological, diplomatic or otherwise.

In 2010, the NPT remains the keystone of global efforts to rid the world of the means of building or acquiring nuclear weapons or other nuclear explosive devices. Emerging during a distant era in which levels of background atmospheric radiation were rising rapidly due to unchecked weapons testing, and in which expectations of the exponential spread of nuclear weapons technology promised a bleak future, the NPT endured despite set-backs *just because* few individuals or states were prepared seriously to contemplate a world without its strictures, and the lines it drew in the nuclear sand. Far more national decision makers eventually came to rely on the certainties, however unsatisfactory in terms of national interests or security, which the NPT provided within an uncertain international system. The depths of the Cold War, far from unhinging these effects, merely reinforced them.

A useful starting point for judging the probability of success for the concept of an international nuclear fuel service is to use the history of the Nuclear Non-proliferation Treaty as a benchmark. On the marginally positive side, the effects of a fuel service of last resort on issues to do with national prestige, state sovereignty or political independence are likely to be rare, transitory and negligible for any state with short-term difficulties in accessing power reactor fuel from commercial sources, *so long as it is in full compliance with its NPT obligations*. Broadly based economic sanctions imposed by the United Nations and extending to imports of nuclear fuel would be extremely unlikely to be implemented unless that state was, in fact, in serious breach of the NPT. In that case, such sanctions would rightly attract widespread support.

More equivocally, the strengthening of international nuclear security which the fuel bank concept would bring about should, on any reasonable interpretation, be seen by states both with and without prospects for using such a service as clearly in their own, individual national interests as a result. Judgments made by individual states on the balance of costs versus benefits to be expected from either "last resort" or permanent fuel bank arrangements will, of course, mandate the reaction of the thirty-five member states of the Board of Governors of the IAEA in accepting for study and development, or ultimately rejecting, the fuel bank proposals of the Nuclear Threat Initiative.

Either way, the world stands at a crossroads which is similar to that confronting it during the mid-to-late 1960s. Then, as now, nuclear technology was threatening to escape irretrievably from an ineffective international security system which, until 1970, had failed to deny to allcomers the means of building nuclear explosives. Then, as now, the world had to deal with atmospheric pollution which threatened dire consequences for the future of the planet. The NPT was the result of overwhelming pragmatism on the part of its non-nuclear founding member states which proved themselves capable of shrugging off national nuclear ambitions in the cause of national (and, of course, international) survival. In 2010, only four states remain outside its membership list, a role-call which, with political will, could be closely paralleled by states participating in the mooted international fuel bank system.

The United States, having in late 2007 supported the NTI initiative with a substantial sum of money was keen, as enthusiasm for the effort gained momentum, to reinforce increasingly positive perceptions around the world about Washington's nuclear fuel security policy thinking. Its framework is provided by the 2004 Global Nuclear Energy Partnership (GNEP) proposal made during the Bush Administration, and reiterated by US Energy Secretary Bodman at the close of the GNEP 2008 Ministerial meeting in Vienna, Austria.

Secretary Bodman pointed out at that meeting that the early work of GNEP member countries had been focused on two distinct elements. One was the need to expand the use of civilian nuclear power for GNEP partners, at all stages of their nuclear power programs and in the context of advances in financing, human resources, safety, and regulatory and legislative frameworks. The second element of GNEP's work, he stated, was the continued emphasis on comprehensive fuel supply and service arrangements, especially at the back end of the nuclear fuel cycle (that is, in the reprocessing of spent reactor fuel or its disposal).

The need to take a fresh look at the development of civilian nuclear energy was underscored, he affirmed, by the filing during the preceding year of fifteen applications with the US Nuclear Regulatory Commission covering the construction of twenty-four nuclear reactors. Finally, Secretary Bodman noted the US Advanced Fuel Cycle initiative which, he claimed, has "... the potential to significantly reduce the volume, thermal output and radiotoxicity of waste requiring permanent, geologic, disposal".[5] At the heart of a program with these aims would be the development of a closed nuclear fuel cycle in which available HEU and plutonium is recovered from spent reactor fuel without the proliferation security issues accompanying earlier reprocessing technologies, and used for further energy extraction in mixed oxide light water reactors.

At the same time, many of the highly toxic products present in spent fuel would be transmuted into less dangerous radio-nuclides. All this would require

5 Closing Statement to the GNEP Ministerial, Remarks prepared for Secretary Bodman, *US Department of Energy, News Archive, Speeches*, October 1, 2008. Available online at: http://www.energy.gov/news/6616.htm. Accessed August 17, 2009.

significant investment by government and industry, and long but indeterminate lead-times. Nevertheless, the Advanced Fuel Cycle proposal (discussed further below) is one indication of American policy movement towards innovative, technology-based ways in which to bring nuclear security under global control through the mechanism of fuel security.

With the active support of thirty-one nations, including the United States, the collective states of the European Union, and other countries in the Middle East and elsewhere, it would be reasonable to expect a cautiously positive reception to the NTI proposal from member states of the International Atomic Energy Agency at meetings of the IAEA Board of Governors. The NTI proposal certainly enjoys an advantage over alternative bids to the extent that it was generated from the initial philanthropy of individuals, rather than the policy preferences and settings of states. As a result, it is reasonable to believe that it would be less likely than similar proposals to suffer from criticism based on the kinds of suspicion over rival state intentions which nation states, perhaps unavoidably, find so difficult to quell. Accusations that the NTI initiative was borne of a desire by the United States and its Western allies to preserve nuclear policy ascendancy, while they will without doubt be made, should on the face of things, be inherently less easy to sustain.

In the event, therefore, it is a matter for serious regret that the Meeting of the IAEA's thirty-five member-state Board of Governors held in Vienna on June 19, 2009 failed to reach consensus on the question of developing a plan for a multilateral civil nuclear fuel supply system of last resort, following the strong recommendations of then Director General ElBaradei. Concerted opposition among some of the Board's developing state members, somewhat stridently led by India, significantly protested "attempts meant to discourage the pursuit of any peaceful nuclear technology on grounds of its alleged 'sensitivity.'"

In contrast to the widespread enthusiasm and high hopes of success, reinforced by expressions of support from Dr ElBaradei, the fuel bank concept met instead with immediate caution and the probable suspicion of relevant developing states concerning the motives of the wealthy states, many of which had supported the NTI fuel bank initiative financially. Concerned for their "inalienable and sovereign rights" under Article IV of the Nuclear Non-proliferation Treaty to develop their own national nuclear fuel cycles, developing states questioned the need for an impartial international source of LEU supply of last resort. In response, the United States countered that the intention of the proposals was indeed to extend, not restrict, access to fissile materials by making them more readily available on a controlled and monitored basis.[6] Once more, trust was found to be thin on the ground in an international forum of immense significance for the human future. At the time of writing, the matter remains before the IAEA Board of Governors for further consideration. This development, while certainly regrettable, does not

6 *IAEA Board splits on nuclear fuel bank proposal*, NTI Global Security Newswire, Friday, June 19, 2009. Available online at: http://www.globalsecuritynewswire.org/gsn/nw_20090619_5943.php. Accessed August 17, 2009.

end the matter, nor the concept of an independently operational global nuclear fuel supply service. It merely postpones its eventual and full consideration by the international community.

The Key Issues

The many elements of traditional strategies for countering state-based nuclear weapons proliferation (such as full scope IAEA inspection and verification programs, and controls over the movement of dual-use technologies through the use of Nuclear Supplier Group export trigger lists) have proved themselves at least partially effective over the six decades of the nuclear era. As opportunities for the covert transfer of fissile materials from power generation to weapons programs continue, however, to tempt some governments, or increase the potential for their supply to terrorist groups, so new and more efficient strategies have become increasingly vital for global security.

The many emergent non-proliferation initiatives, with the prominent contribution of the nuclear fuel bank concept, work both to deter such states from acting as proxies for the ultimate outrage of nuclear terrorism, and to deny operational nuclear fission devices to terrorist groups willing to act independently of sympathetic governments. Many have been discussed earlier in this analysis, but it opportune here to reiterate the scope and nature of the challenges involved in denying further access to nuclear explosives to all agents into the future. At the same time, it remains incumbent on all those states holding stocks of nuclear weapons to take concrete and time-limited steps to divest themselves completely of these essentially useless devices.

A great deal of weapons-grade fissile material will, nevertheless and without doubt, remain available for both legitimate use and unacceptable misuse around the world over the long term, with no knowable time frame for its complete elimination. All told, there is still some 1,900 tons of highly enriched uranium in the holdings of over thirty countries around the world, of which around 100 tons is in civilian hands (and is itself sufficient for perhaps two thousand nuclear weapons). HEU and plutonium are together stored in over forty countries.

This quantum and diversity will, despite efforts to the contrary, continue to exist essentially unchanged for many years in stocks of fissile material held in the national reserves and operational weapon arsenals of the five declared Nuclear Weapon States, as well as the much smaller but significant holdings of India, Pakistan, Israel and North Korea. Furthermore, HEU will also continue to find employment in relatively small quantities for specialized, non-nuclear weapons purposes such as, for example, military marine propulsion for nuclear powered submarines and ice breakers. It will also continue to be used as targets in the production of medical and other isotopes (for example, the diagnostic use of molybdenum-99) and in a number of research reactors and other facilities around the world.

Many of the civilian facilities now holding HEU stocks, such as the many university research reactors in the United States, Russia and elsewhere, still do not measure up to the status of military holdings in terms of the physical security regime in force to protect against theft by criminals or terrorists. The South African Pelindaba incident noted earlier is a timely reminder, if one were needed, of the capacity for potentially catastrophic consequences in the event of security breaches leading to the theft of HEU or plutonium, and its subsequent use in terrorist acts.

Given the size and nature of the threat posed by insecure nuclear materials, and even operational weapons, around the world, the United States has undertaken a number of policy initiatives. One significant initiative has been the 2006 joint US-Russia Global Initiative to Combat Nuclear Terrorism, announced by Presidents Bush and Putin at their St. Petersburg Summit Meeting of that year. Added to this is the two presidents' Bratislava accord on nuclear security of the previous year, as well as the Global Threat Reduction Initiative (GTRI) of 2004.[7]

These three ambitious international policy moves by the United States and Russia are clear indicators of the urgent and cooperative spirit in which the two countries view the dangers of nuclear based terrorist threats, and those posed by nuclear aspirant states. The 2004 GTRI joint initiative on combating the global threat posed by vulnerable nuclear facilities uses a triad of approaches. First, converting existing research reactors and isotope production facilities world wide from HEU to LEU fuel use eliminates weapons-usable fissile materials from the civilian fuel cycle. That program will conclude in 2014. Second, it removes surplus weapons-grade material of Russian, US and other origin (including the large quantities of US "Atoms for Peace" era legacy material) as well as vulnerable radiological material usable in "dirty bombs," from civilian nuclear sites around the world. That effort continues, and will conclude in 2019. Third, the program establishes adequate interim containment and protection standards in the many nuclear facilities in which inadequate physical security measures, especially in the former Soviet sphere, have resulted in an unnecessarily large potential for material theft or sabotage.

One year later, the 2005 Bush–Putin Bratislava Summit accords laid the foundations for more comprehensive anti-nuclear terrorism measures by re-establishing the commitment of both sides to accelerating delayed security upgrades, at least to minimally acceptable international standards, then awaiting installation at the many inadequately protected Russian nuclear sites.

By 2009, the 2006 Global Initiative to Combat Nuclear Terrorism, the product of the efforts of thirteen countries meeting at Rabat, Morocco, numbered some seventy-six states linked by a common commitment to combat global nuclear terrorism in a determined and systematic, rather than uncoordinated or ad hoc, way. Although not a legally binding international arrangement, the Global Initiative

---

7 Nuclear Threat Initiative, *Securing the bomb: A summary of the key issues*, 2007. Available online at: http://www.nti.org/e_research/cnwm/overview/issue.asp. Accessed September 4, 2009.

seeks to enhance information and expertise sharing between member states in practical and cooperative ways.[8] Unlike those nuclear fuel security initiatives which are largely confined to United States–Russia bilateral cooperation, the Global Initiative to Combat Nuclear Terrorism has been formulated to provide for the largest possible level of participation by states, in the belief that solutions to this global security crisis must ultimately be global in their origins and scope. It remains to be seen whether the low key but vitally important activities embedded in this initiative, such as putting in place better accounting, control and physical protection of nuclear materials, as well as strengthening operational forensic detection capabilities, will provide reliably enhanced levels of security for its member states. At the very least, though, the program demonstrates that significant levels of cooperation in the securing of vulnerable nuclear materials can be accomplished when the perceived need is sufficiently broadly based, serious and pressing.

Together, these three programs amount to a promising beginning to the task which confronts the United States in repatriating and gathering securely together the weapons-grade fissile material which has accumulated in civilian hands, now for decades. The facilities which use this material for legitimate research purposes no longer require HEU, and they and their governments can usually be convinced of the security benefits to be gained by conversion of reactors and critical research assemblies to low enriched uranium. The success thus far gained demonstrates that even so complex a task as the impermeable quarantining of existing HEU and plutonium in world stocks is capable of reaching current (though admittedly indistinct) IAEA fissile material security standards, and to do so on a durable and comprehensive international basis. Further advances in the cooperative application of universal international security standards largely await the political will of the international society of states, and must constitute a prerequisite to long-term nuclear material security goals.

Nevertheless, even with growing recognition of the size and nature of the threat of nuclear terrorism, together with traditionally understood modes of state-based nuclear weapons proliferation, the eternal dilemma for those who defend their ground against external attack remains: their attackers may fail many times and yet continue to mount new attacks, while the defenders will only fail once, usually with catastrophic results. This reality lies at the center of counter-nuclear weapon proliferation and anti-terrorist policy planning, and drives the multi-faceted nature of efforts to deny success to those states or non-state groups which, once in possession of nuclear explosives, would surely rush to use them against Western targets, most likely beginning with the United States itself. Earlier discussions within this analysis have made this reality abundantly clear. Furthermore, fully effective border control which is reliant on forensic detection and interdiction of nuclear devices moving across America's immensely long international boundaries

---

8 US Department of State, *The global initiative to combat nuclear terrorism*, 2009. Available online at: http://www.state.gov/t/isn/c18406.htm. Accessed September 4, 2009.

is probably impossible. Conversely, it may well be an enemy located undetected within the United States itself, rather than external agents, which poses the greatest threat to homeland security, as was the case in the terrorist attacks of September 11, 2001.

Much, then, has been accomplished within the limits defined by the combined efforts of the United States, Russia and the many states committed to working together to secure the world's nuclear future. But what of the few states with nuclear weapons (or aspirations to acquire them) and no intention of divestiture? Of these, Pakistan and North Korea, with their histories of political instability, totalitarian unpredictability, Islamic zealotry, official corruption and demonstrated preparedness to sell nuclear weapons technologies to terrorist groups and aspirant states, present the most serious and durable threats to the world's hopes for nuclear security. The reality of over forty domestic and trans-national terrorist groups operating within Pakistan, often with links to Pakistan's own intelligence agencies, to al-Qaeda and its affiliates, or to Afghan Taliban forces, only underlines the extreme urgency attached to securing Pakistan's nuclear establishments, fissile materials and nuclear weapons systems.

Meanwhile, the de-commissioning of former Soviet nuclear weapons, and the removal of more than 200 tons of their fissile material for down-blending and use in US nuclear power plants continues under the Nunn-Lugar Cooperative Threat Reduction Program (CTR). This initiative, which dates back to the legislative efforts of Senators Sam Nunn (GA-D) and Richard Lugar (IN-R) in 1991, has resulted in the de-activation of some 7,500 Russian and former Soviet era nuclear warheads, and the upgrading of security at twenty-four Russian nuclear weapon storage sites.[9] Each of these accomplishments, and several other achievements related to intercontinental ballistic missile elimination, are examples of the potential of the efforts of well-placed and committed individuals working in the global strategic security field. They echo the achievements thus far of the Nuclear Threat Initiative discussed earlier.

Several additional key issues complete the picture of nuclear fuel security used as the s*ine qua non* of non-proliferation and nuclear terrorism policy planning on the national and global scales. Predominantly aimed at disallowing nuclear explosives in the hands of terrorists, they together provide further comfort for those who take a position in favor of the potential of international political cooperation when allied to the ever-growing detection and interdiction capabilities of current and future technologies, and the ingenuity of the world's diplomatic and scientific resources.

One issue of long-term significance (and one which extends beyond non-proliferation imperatives alone) is the development of increasingly proliferation resistant reactor technologies which do not allow for the separation of plutonium

9 Richard G. Lugar, United States Senator for Indiana (R), *Nunn – Lugar cooperative threat reduction program*, 2009. Available online at: http://lugar.senate.gov/nunnlugar/index.cfm. Accessed September 5, 2009.

from spent fuel, nor leave significant quantities of HEU available for use in fission weapons. A dedicated discussion of the complex history and current state of development of the world's various reactor technologies, their generational structure and inherent strengths and weaknesses, is beyond the scope and limits of this analysis. It is sufficient here, however, merely to characterize the four generations of full scale energy production reactor technology in order to establish the importance of the emerging Generation IV technologies, their status and potential to resist weapons proliferation.[10]

The first generation of the world's nuclear reactors, dating back to the 1950s and 1960s, are now confined operationally to the United Kingdom, and these are in the process of being retired. Most of the current fleet of US production reactors, and most nuclear power now generated in the rest of the world, use Generation II technologies. Comprising a number of discrete types, but typically derived from earlier naval reactor designs, they are safe and reliable while being inherently more complex and expensive to build and operate than more recent designs. Significantly, they also produce greater quantities of unburned fuel and radioactive waste materials than successor technologies.

Numbers of new and more efficient designs are, however, now beginning to come on stream from the world's suppliers of nuclear reactors. These Generation III and 3+ Advanced Reactor designs, originating in the United States and Japan among others, are usually larger than their predecessors and often involve international technical collaboration. They feature standardized, more rugged designs and reduced operating costs, while burning greater proportions of their fissile fuel load than earlier technologies.

For the longer term future, however, the United States has brought together the Generation IV International Forum (GIF) of thirteen participating nations, which was set up in 2001 to evaluate six new reactor concepts for possible development and deployment between 2010 and 2030.[11] These countries, dominated financially by the United States, France and Japan, believe that the six concepts for evaluation represent the future of nuclear energy in that they all promise advances in sustainability, economics, safety, reliability and proliferation resistance. Research project expenditures to investigate the viability of the six reactor concepts stand at some $US6 billion over fifteen years. Most reactor technologies to be investigated feature closed cycle architectures to maximize fuel usage while minimizing their output of spent fuel in the form of high radiation level wastes needing long-term geological sequestration.

Whether any or some of the new Generation IV reactor technologies, with their promise of progress on several fronts, eventually prove commercially scalable

10 World Nuclear Association, *Advanced nuclear power reactors*, July, 2009. Available online at: http://www.world-nuclear.org/info/inf08.html. Accessed September 6, 2009.

11 World Nuclear Association, Generation IV nuclear reactors, 2008. Available online at: http://www.world-nuclear.org/info/inf77.html. Accessed September 6, 2009.

and viable in the long term is not knowable at this stage. Nevertheless, they each confirm the power of science and innovation to challenge and resist the most pernicious use of technology for applications which threaten the security of the world's people.

Another key issue for the United States in its homeland security planning is how best to use available resources in detecting and interdicting nuclear materials being smuggled both across US borders and around the wider world. The successful prevention of nuclear smuggling, as a last line of defense when radiological or fissionable materials have already been stolen from insecure nuclear facilities, is extremely difficult to accomplish. One of the most intractable obstacles is the fact that non-irradiated (or "fresh") highly enriched uranium and plutonium are only mildly radioactive, and therefore difficult to detect using remote sensing technologies. Once again, the eternal dilemma of the defender, mentioned earlier, comes to the fore.

Even so, 80 or more states now support the 2003 Proliferation Security Initiative of the United States, which anticipated and responds to United Nations Security Council Resolution (UNSCR) 1540 of 2004, establishing binding obligations on all UN member states, under Chapter VII of the UN Charter, to:

> … refrain from providing any form of support to non-State actors that attempt to develop, acquire, manufacture, possess, transport, transfer or use nuclear, chemical or biological weapons and their means of delivery….[12]

This United Nations move is a necessary and welcome addition to the arsenal of enforceable international law instruments mandating positive and negative actions by states in respect of the proliferation of nuclear weapons technologies and materials into the hands of individuals and non-state groups who should never acquire them.

Moreover, the US is currently undertaking programs such as the Department of Energy's Second Line of Defense initiative aimed at installing equipment capable of detecting nuclear materials at several hundred vulnerable choke points in Russia, as well as a number of other states in Eastern Europe and the former republics of the Soviet Union. A further program is the Department of Energy's Megaports Initiative, which aims to install radiation detection equipment at some seventy-five of the largest ports around the world from which the bulk of international shipping departs for the United States. These two initiatives represent only a sample among several programs developed by DOE, the Department of Homeland Security and the Bureau of Customs and Border Protection and focused on issues such

12 United Nations Security Council Press Release SC/8076, *Security Council decides all states shall act to prevent proliferation of mass destruction weapons*, April 28, 2004. Available online at: http://www.un.org/News/Press/docs/2004/sc8076.doc.htm. Accessed September 8, 2009.

as security enhancement for shipping container and other freight types, and the equipping and training of overseas border protection officials.

Once again, however, the commitment of resources and personnel will only ever provide a less than perfect solution to the nuclear materials security conundrum. As in the case of illicit smuggling of other kinds of contraband, such as narcotics and small arms, the challenge of fissile and radioactive materials border control for the United States is unceasing, in a state of constant flux and demanding of unremitting vigilance.

Conclusion

Each of the key issues in nuclear fuel security which have been canvassed in this chapter hold clear potential to significantly enhance prospects for bringing nuclear weapons proliferation, and in particular the acquisition of nuclear weapons by terrorists, under more coordinated and strengthened control. Just as clearly, though, there is an obvious but crucial nexus between state and non-state based processes of nuclear diffusion which require policy advisers and their governments to take a holistic, coordinated and pragmatic approach to prevention, detection and interdiction strategies.

Plainly put, it is neither acceptable nor sufficient for the governments of states to move forward on nuclear non-proliferation policies in ways which will, or may, tend to hamper the capacity of other states (either individually or in alliance) and of the United Nations to evolve effective universal counter-measures. The reported reaction of some states as members of the IAEA Board of Governors, in response to the Nuclear Threat Initiative's nuclear fuel bank proposals discussed earlier, is a clear example of this phenomenon in action.

A far more effective and generalizable approach in this case would have been for those states with doubts as to the efficacy or security implications of such a move to acknowledge that their own individual national security priorities are best served by the application of universal solutions to global crises in the arena of nuclear proliferation. Alone, each recalcitrant state can go only so far in, for example, preventing widespread nuclear smuggling and the prospect of nuclear terrorism. At best, this would amount to a regional and incomplete effort. In concert with the broad society of states, however, and with the functional catalyst of the International Atomic Energy Agency in support, much more progress in nuclear non-proliferation and counter-nuclear terrorism clearly becomes possible.

States which insist that fully independent nuclear power generation capabilities are a matter of sovereignty, international prestige and the invulnerability of national security are being increasingly confronted by the need to convince the society of states that such is the case. They must be prepared to build a very convincing case, backed up by the transparency of IAEA-conducted full-scope inspection and verification activities, in denying (as Iran in 2009 continued to insist) any intention to seek to divert nuclear fuel for use in power generation towards a weapons program.

Finally, the fuel-based approach to nuclear security and non-proliferation, for both the state and non-state actor cases, is possibly the single most effective way in which international efforts to quarantine the spread of nuclear explosives technologies can flush out the real aims of states suspected of harboring aspirations (confirmed in the case of Pakistan) not only to acquire nuclear weapons but to spread their technologies more widely.

The exclusively domestic manufacture of nuclear fuel ready for loading into a power production reactor is both complex and costly, as is its re-processing to recover unused fuel, and the eventual long term sequestration of highly radioactive waste products. The various stages of the nuclear fuel cycle, from uranium exploration, mining and milling to gas conversion and centrifuge cascade enrichment, and on to fuel fabrication, short term storage, possible re-processing and long term sequestration of waste materials are simply not a practical or economic option for many smaller states.

If cost and technological complexity were not enough, the question of access to a reliable supply of uranium oxide at the very beginning of the nuclear fuel cycle is also problematic. A state may simply not have sufficient indigenous uranium resources to fuel its planned nuclear power assets; few states are so well endowed. It is here, then, that nuclear fuel exercises its enormous political leverage. Should a state show without doubt that it is not intent on using low enriched uranium for a*ny* purpose not sanctioned by the international community, it will not only enjoy reliable and economically viable access to the enormous and essentially environmentally benign energy of nuclear fission, but also the avoidance of a need to process and store on its own territory the resultant radioactive waste products.

The test of a state's nuclear good faith which this represents ought to be sufficient evidence with which to judge the intentions of any state in a world in which nuclear weapons proliferation threatens all states and all people. The United States must calibrate its own future nuclear policy and security settings with this reality in mind.

Chapter 9
The Future: Alternative Policy Pathways for the Twenty-first Century

Introduction

Thus far in this analysis, emphasis has been placed on current and near-term strategies for nuclear fuel security. The penultimate chapter will turn to consider more speculative and slowly evolving concepts and approaches which appear to warrant our attention in view of their promise of progress built on previous success, as well as reappraisals of traditional non-proliferation architectures which now carry renewed relevance, and innovative or untested strategies whose time has come.

The international context for this exercise is one in which the regional and global uncertainties and security dilemmas which resolved into sharp focus during the first decade of the twenty-first century have now coalesced around a wider acceptance in the nation states of the need to cooperate more closely, on both the regional and global scales, in confronting existential threats ranging from climate change to nuclear war. There is a growing recognition at all levels of public life, in government circles and throughout civil societies that, in the future, those things which bring the world's people and potentialities more closely together will prove to be even more seminally important and durable than those which separate and divide us. There is also a strong belief that this is so despite (or perhaps *just because* of) the existence in the world of politically, religiously and culturally driven forces which have declared total war on modernity and the liberal or Western political tradition.

Much work in the cause of nuclear fuel security and non-proliferation remains to be done, many elements of which must now be accomplished in short order if the international system is merely to hold the line against the accelerating diffusion of nuclear know-how, capacities and intent. This will require innovative reappraisals of many closely held nostrums and fiercely defended assumptions embedded in the status quo of the non-proliferation policy world. These must come from those in, and close to, government, from the academy, from civil society, and from specialized international non-governmental and governmental organizations of many kinds. Such a reappraisal will also demand a thorough reconsideration of the way forward for the international law of disarmament, effective nuclear non-proliferation and the long term security of all kinds of fissile and radioactive materials. But most of all, though, it will require what is often in tragically short supply in international politics – a spirit of cooperation, trust and moral courage.

At the risk of accusations of speculative over-simplification and even idealism, it should be possible at this point to review a range of structural elements which will surely form the backbone, and background, for nuclear security success which is based on fissile material security. These building blocks – the various non-proliferation strategies, foci, diplomacies, technologies, international rules (and rules about making rules), regimes of common understanding, as well as the work of international governmental and non-governmental organizations, and many more – are well enough understood in their current roles. They may not be so completely comprehended in terms of their present and future places in the wider whole of nuclear security, their capacity for adaptation, interpretation and reinterpretation, nor the need for their realignment or for renewed relevance in their application to the nuclear security goals sought. Furthermore, new nuclear security measures, new ways of closing the open circle of threat, must be created, recognized for what they are, and imagined for what they may become.

In other words, it is high time in 2010 for a new, more integrated and closely coordinated global effort to secure the world's nuclear future. In undertaking this task, the United States, its allies and friends, and those other states whose peoples and governments truly seek nuclear security, will also secure their own futures. The first, but vital, step in this long journey is to start to imagine what nuclear success would look like. The second step is to begin to image, at least in outline, the journey which ends at that point. There is much to be done and time is pressing.

What, then, would an acceptable version of world-wide nuclear security look like? What would it encompass, in what would it consist, and what would be its necessary qualities, capacities and nature?

Imagining Nuclear Security Success

A future nuclear security success story would present the world with an integrated and many-sided matrix of mutually supportive conceptual and operational architectures, combined with universal and enforceable rule sets, agreed procedural methodologies, closely targeted and monitored policy implementation and enforcement machineries, and universally coordinated emphasis on fair and just outcomes which are sustainable, in all senses of the word, into the distant future.

Nuclear security success will not be achieved through ad hoc or piecemeal approaches aimed at denying access to fissile materials, expertise or technologies for nuclear-aspirant agents. This is tantamount to deciding that treating the symptoms of a malaise is more important or necessary than addressing the source of the disease.

Nuclear security can, however, be achieved when the benefits to the security and prosperity of the nations states, and the welfare of their people (which are readily and transparently available to states not seeking nuclear explosives capacities) clearly and unequivocally outweigh their acquisition. When this is easily and unambiguously demonstrable in the international public domain, and an effective global nuclear security system is fully capable of exhibiting to the world the real intent (whether benign or malign) behind those countries suspected of seeking nuclear weapons, states will eventually be prepared to rely on fully coordinated and globally effective non-proliferation capabilities sufficient to allow predictable and proportionate responses at the United Nations Security Council level to take their course.

Those states willing to let such an operational system work as designed will, all else being equal, also tend to be more prepared than in the past to participate in international action to halt further development, acquisition or use of nuclear weapons, their delivery systems or the fissile material without which they pose no immediate threat. While it will certainly take some time for general, and eventually universal, trust in a global enforcement regime of this nature to coalesce, nevertheless, early or moderate successes will begin to build a broader, more reliable foundation on which to construct a more comprehensive non-proliferation successor regime.

The most important aspect of any human artifact is the strength of its foundations which, in this case, must begin, as introduced earlier, with the international legal context. The Law of Nations suffers from a lack of popular, and at times even specialist, understanding of its nature, capabilities and applications. In the case of the world's nuclear affairs, the international law of conflict, humanitarianism and human rights has much to say concerning the nature of nuclear war, the uses of fissionable materials, their international trade, transportation and physical security, nuclear waste management standards, international materials diversion safeguards, and the prevention of nuclear terrorism – among many other nuclear issues of surpassing international importance.

There is insufficient space in these pages to discuss issues beyond the potential of international law to reinforce and support the non-proliferation and nuclear counter-terrorism policies and actions of the nation states and the United Nations, nor to provide any but the clearest examples of international law in action in the cause of nuclear non-proliferation arrangements. Nonetheless, international law most assuredly matters to the world and to the international society of states and their governments because (first and foremost) the states *say* that it does. International law brings the imprimatur and cachet of *law* to the policies and decisions of states exercised through their governments.

In this way, agreements which are perceived to require the force of law become binding international treaties; long term customary relationships eventually take on the force of legal obligations voluntarily undertaken, while vitally important universal agreements (such as the 1970 Nuclear Non-proliferation Treaty) become multilateral United Nations Treaties whose strictures and intent are acknowledged universally and flouted at considerable national cost. When an international arrangement or agreement must not be allowed to fail, international law attains its greatest relevance and strength.

Why so? Although there are few avenues available for the enforcement of rights claimed and obligations owed in international law (this is slowly changing, especially in respect of individual rights) states usually feel constrained to toe the line marked by the extent of their current legal rights and obligations for fear that failure to do so may severely curtail their future credibility, and thus possibly deny them access to desirable international rights and advantageous relationships yet to be offered.

The NPT is, of course, an archetype for durable international legal agreements, and one which, despite its shortcomings, continues to lie at the heart of all international nuclear affairs. For forty years it has set the global nuclear agenda, established the form and structure of the primary international nuclear bargain made during the mid nineteen-sixties in the use of nuclear energy, and become, in the process, seemingly irreplaceable. It is the apotheosis, as relevant as any exemplar in other spheres of international life, of international law as the self-enforcing articulation of actions taken in response to perceived urgent need. Its durability and stability is legion, and, to repeat an earlier point, unlike its early and later interpretation and application, its text has never been altered in any way whatsoever. In this case, stability in the interpretation of international treaty law has been universally regarded, rightly or wrongly, as superior to plasticity of content over time.

On this account, then, the world sorely needs international law in order to help nuclear non-proliferation initiatives, such as the current international fuel bank proposals, sustain challenges made against them, and more so now than at any time in living memory. Many unresolved initiatives discussed during this analysis would certainly benefit from their being infused with international legal obligation. Among them is the long sought, but highly elusive multilateral Fissile Material Cut-off Treaty (FMCT) concept, which was framed to prohibit all new production of highly enriched uranium and weapons grade plutonium. Should the proposal for an IAEA-led international nuclear fuel bank eventually succeed, such an arrangement will necessarily be reflected by a multilateral United Nations treaty laying out the rights and obligations of states, while guaranteeing the uninterrupted supply of uranium for nuclear energy production in last resort circumstances. Without such an international agreement, no proposal of this kind, or size, can realistically be expected to succeed.

Many other vital multilateral nuclear agreements are reflected in international law, from the 1996 Comprehensive Test Ban Treaty (still not in force in 2010) to the many bilateral Safeguards Agreements made between individual states and the International Atomic Energy Agency, and the 1994 Convention on Nuclear Safety, all of which carry the full force and power of international law. They each, along with many others, establish the existence and scope of legal obligations and rights for states as members of international society. It is therefore no exaggeration to assert that nuclear international relations, in all their complexity, are dependent for their continued existence on the growing strength of the force of international law.

Practical, flexible, durable and self-enforcing international law is, of course, a necessary but insufficient guarantor of nuclear security success. The most obvious thread running through the broad characterization, noted above, of what a future nuclear security success story should look like is probably the conceptual and operational cooperation without which the current deficits in nuclear security will not be addressed successfully. The nature of international affairs being what they are, and as the history of the United Nations so clearly reinforces, it is an unfortunate reality that any international confirmation of something approaching a consensus on the way forward for nuclear non-proliferation and nuclear fuel security is often seen as either politically and pragmatically impossible, or at best a fool's errand. Anything approaching the nirvana of the "universal and enforceable rule sets, agreed procedural methodologies, closely targeted and monitored policy implementation and enforcement machineries" mentioned earlier appears, on this account, to lie within these two grim categories. The hostile reaction of the developing states at the 2009 meeting of IAEA Board of Governors, called to consider the Nuclear Threat Initiative's nuclear fuel bank proposals (a protest led by India) is symptomatic of views in this genre.

Such a negative interpretation should not, in any way, deter states which understand the potential dangers inherent in nuclear proliferation and nuclear terrorism from building regionally or functionally limited cooperative relationships aimed at countering such hostility. However, new groupings or alignments of states formed with this intention must use the opportunity to fix in place the non-proliferation initiatives which remain insufficiently well supported, through redoubled diplomatic effort at the IAEA and the many formal and less formal international governmental organizations with duties in the nuclear arms control field.[1]

This amounts, in principle, to a move away from a tendency towards global coercion, much of it emanating historically from the Nuclear Weapon States, their friends and allies, and towards diplomatic engagement and cooperative exchanges of views, ideas and conceptual options for policy change. Progress in the control of nuclear diffusion, including through nuclear fuel security enhancement, is likely to be more effective and long-lasting when it is predicated on collective, rather than unilateral action, on dialogue and integration rather than isolation and confinement.

The ruling principle, as the nuclear non-proliferation regime continues slowly to erode, in denying nuclear devices to regimes and terrorists with malign intentions must, therefore, be "trust but verify." As a watchword for the evolution of regionally based and focused coalitions of states on nuclear policy affairs, led by the United States, its allies and others, this aphorism neatly encompasses the end-point for states seeking to avoid their international legal obligations on nuclear materials diversion. However, regional coalitions of states brought together to

1 Examples of organizations in these categories include the Nuclear Suppliers Group, the Proliferation Security Initiative, the UN Conference on Disarmament, the OECD Nuclear Energy Agency, and many more.

resolve nuclear issues on, for example, the Korean peninsula, the Indian subcontinent or the Middle East must bring something new to the non-proliferation table if they are to succeed.

There is no shortage of worthwhile initiatives, beginning, of course, with the notion of international nuclear fuel bank (and perhaps re-processing) services. While this is itself a difficult diplomatic proposition, it is not alone. Another intractable concern will continue to be the problem of the illicit trafficking and authorized (but vulnerable) movement, usually across international borders, of fissile materials and – even more difficult to control – the expertise and knowledge necessary to build a weapon of mass destruction. Furthermore, each of these three mooted regional nuclear non-proliferation targets contains its own unique strategic security, diplomatic, political, social and economic problems and concerns. Each one, however, must be addressed within a globally relevant framework which accepts the universality of nuclear challenge and response. A summary enumeration of a selection of only the most urgently needed responses to the gathering nuclear proliferation and terrorist stormclouds lends scale to their scope and dimensions:

- Accelerating international cooperative work on Generation IV reactor technologies, with emphasis on those promising significant progress in proliferation resistance (including plutonium non-separation).
- Securing and consolidating world-wide civilian fissile materials stocks of highly enriched uranium, and some 250 tons of "reactor-grade" plutonium separated for, but not used in, fission weapons during the Cold War, with the aim of eventually eliminating them.
- Completing the transfer back to United States control of superfluous highly enriched uranium currently remaining in use or storage in conjunction with research reactors and critical research assemblies worldwide.
- Completing the down-blending and transfer to the United States of up to 500 tons of Russian former weapons-grade HEU. The material should ultimately be sequestered under IAEA control until its final use, in blended-down form, as reactor fuel for energy production in the United States.
- Meaningful progress and ultimate success in strengthening the scope of relevant international law and safeguard monitoring of plutonium and HEU production in the Nuclear Weapon States. This should be achieved by both a multilateral Fissile Materials Cut-off Treaty (FMCT) prohibiting new production of HEU and plutonium, and a comprehensive United Nations Nuclear Weapons Convention verifiably prohibiting the development, acquisition or deployment of nuclear weapons and devices, their materials and technologies (in a form similar to the present Chemical and Biological Weapons Conventions).
- Implementing United Nations Security Council Resolution 1540, which creates legal obligations for all states in areas such as appropriate and effective nuclear security through criminalizing nuclear theft and terrorism,

and putting in place effective border controls to detect and interdict nuclear related shipments crossing international boundaries.

- Consolidating hundreds of tons of fissile material in the US nuclear complex in fewer locations, while significantly strengthening its physical security.
- New and deep cuts to the numbers of warheads in the deployed arsenals of the United States and Russia, beyond the current 2003 Moscow (SORT) Treaty levels, and perhaps as low as 1,000 warheads each in a successor treaty to START I, which expired in December, 2009. Only by such an initiative will the international political impetus for effective nuclear proliferation and nuclear counter-terrorism measures undertaken in the dominant nuclear states reach sustainable levels of credibility and universal support.
- Bringing together in more effective, efficient and complementary ways the oversight, trigger listings and export control work of the Nuclear Suppliers Group, the Zangger Committee and the Wassenaar Arrangement in respect of nuclear and dual-use technology exporting countries.
- Ensuring that the 1970 Nuclear Non-proliferation Treaty is reinforced as far as possible through the successful conclusion of its five yearly Review Conferences. The next conference, to be held in New York during May, 2010, will be crucially important for continued progress in all aspects of nuclear non-proliferation.
- Learning from the A.Q. Khan nuclear smuggling experience in making further progress on the detection and interdiction of nuclear smuggling activities worldwide, with increased resources allocated to programs such as the US led Proliferation Security Initiative and US and other states' assistance to the border control authorities of developing states.
- Further emphasis within the Nuclear Weapon States, as well as other economically and technically advanced member states of the Western Alliance, on the provision of technical and other assistance to developing states believed vulnerable as potential targets of nuclear terrorists seeking safe haven territorial bases.
- Domestically, the United States must redouble its nuclear detection and interdiction efforts at its international boundaries and ports, at points of departure of incoming shipments, and at all vulnerable points within the territory of the United States.

Only two of these broad strategies are in any sense departures in international nuclear non-proliferation policy settings: the international nuclear fuel bank or service and the counter-proliferation promise of Generation IV reactor technologies. A third, involving the guaranteed international supply of low enriched uranium to states on the basis of agreed levels of installed nuclear power generating capacity, will be discussed in the final chapter.

These and other measures, if pursued with vigor, with effective use of appropriate levels of funding and other resources, and with the unimpeded cooperation and contribution of the US, its allies and friends, and states not aligned with the Western Alliance, will not result in immediate or rapid cessation or de-escalation of risk to United States homeland security. They will, however, give the US and the world a breathing space sufficiently wide to allow for the introduction of a step change in the coordination of countermeasures against nuclear terrorism, and proliferation more generally understood.

Once achieved, it is essential that the momentum generated in this way is not lost in what may well be the single most pernicious of challenges in the nuclear non-proliferation field – that of complacency. There remains no doubt in 2010 that terrorist groups of many stripes, and from a range of cultural and political origins, continue to strive for their nirvana in the shape of a workable improvised nuclear explosive device (IND), or even a sophisticated nuclear fission weapon. There is also no plausible doubt that, once acquired, they would detonate such a device with the aim in view of creating mass casualties, numbering in the many thousands, in a large and densely populated city. No warning or demonstration of intent of the kind proposed but not attempted by the United States towards Japan in the closing days of the Second World War can be anticipated. No regret for the catastrophic consequences of failure to act to thwart a known and specific risk of nuclear terrorism could redeem those who were in a position to do so.

The introduction to this chapter foreshadowed a three-part analysis of the way forward, comprising more speculative approaches to nuclear fuel security. It noted the need to add to redoubled efforts with strong records of success by addressing strategies holding out the promise of progress built on reappraisals of traditional non-proliferation architectures which now carry renewed relevance. It also pointed to truly new and innovative or untested strategies as a necessary part of the nuclear security mix.

A Reappraisal

Thus far in the history of the nuclear fuel cycle, security of nuclear fuel has been, at best, inconsistent and, at worst, dangerously lax. This is so not only in the political sense in which policy change has failed to track developing and growing threat levels in specific security scenarios (such as the physical consolidation of the US nuclear stockpile). It is also true of the need for renewal and enhancement of the technical basis for the success of policy initiatives to secure nuclear materials and technology.

If international control over fuel security is to count for more than it currently does, and do so consistently and on a global basis, then fissile materials accounting, inspection and verification activities must be significantly enhanced in respect of both the nuclear and non-nuclear weapon states. This heavy burden of responsibility bears primarily, of course, upon the International Atomic

Energy Agency, based in Vienna, Austria. Under its retiring Director General, Dr Mohamed ElBaradei, the IAEA has been extremely efficient in its employment of resources which, unfortunately, remain scarcely sufficient for the minimally acceptable discharge of its statutory responsibilities. The introduction, during the early 1990s, of full scope safeguards against the illicit transfer of fissile material from civilian to military programs through the negotiation of bilateral Additional Protocols has further strained the Agency's coffers and technical and personnel resources through the need for increased numbers of more extensive inspections of nuclear facilities. The inhibited capacity of the IAEA to fulfill its rapidly expanding remit, as many new nuclear power plants are planned and constructed in both existing and newly nuclear states, means that a failure by its member states to properly fund the agency's activities represents a clear and growing danger of the undetected misuse of fissile material. Their current failure adequately to support the world's only comprehensive nuclear international agency must be corrected before its consequences materially threaten the full range of global nuclear security settings.

The IAEA and its future is seminally important in the chances for success both of traditional strategies requiring renewed policy action, and non-traditional and innovative approaches to non-proliferation efforts. As Dr ElBaradei stated in August 2009, the ability of the IAEA's independent nuclear verification system to detect the possible diversion of nuclear material and undeclared nuclear activities is already in jeopardy. The agency's analytical laboratory assets are, he emphasized, outdated, while it lacks sufficient access to satellite imagery as a means of detection of clandestine nuclear programs. Without a sufficient increase in funding, this situation would only deteriorate. Furthermore, a dramatic staff shortage in its nuclear safety program is opening up the real possibility that a Chernobyl-type disaster could occur again.

ElBaradei added that the risk of terrorists obtaining nuclear or radioactive materials, and using them in a potentially devastating attack, will grow if the highly inadequate staffing and funding under which the IAEA now labors is not dealt with. As a parting shot, he noted that the agency required an immediate infusion of 80 million euros to rebuild its dilapidated infrastructure.[2] This unusually frank admission of the inadequacy of the UN nuclear watchdog's resources and capacities was made politically possible only by the impending departure of its highly regarded and long-serving Director General. Whether his impassioned plea for sufficiency of funding will be heeded by the IAEA's member states, led by the United States, remains uncertain at the time of writing, but must remain a prerequisite for all hopes of reversing the steady decline of the agency's capabilities, and with that, of the goal of sustainable nuclear fuel security.

---

2 UN atomic energy agency's board gathers to discuss upcoming budget, *UN News Centre*, August 3, 2009. Available online at: http://www.un.org/apps/news/story.asp?NewsID=31658andCr=iaeaandCr1=. Accessed September 19, 2009.

If this last analysis represents the nadir for optimism over future security prospects, it must be observed that change is the single certainty of international affairs, as it is of all human activities. In this instance, however, renewed focus on progress is time-limited. Where, we must ask, can the international society of states find new ways to tread old pathways within the non-proliferation paradigm in enhancing their capacity to forestall a nuclear fuel security break-out?

One clear opportunity lies in one of the most resented of all the obligations which the 1970 Nuclear Non-proliferation Treaty imposes on non-nuclear weapon states. Article III of the treaty exempts the nuclear weapon states (that is, the five states which had manufactured and exploded a nuclear weapon or other explosive nuclear device before January 1, 1967) from the safeguards inspection and verification regime which had already been undertaken by the IAEA prior to the treaty's entry into force. Although the five nuclear weapon states – the United States, the Soviet Union/Russia, the United Kingdom, France and China – each offered to undergo limited "voluntary" inspections of their civilian (but not military) nuclear facilities, the IAEA has only ever been able to inspect a small minority of their many nuclear sites.

Should the five NPT-acknowledged nuclear weapon states end this arrangement, which in the NPT era has always been controversial and a source of discontent among many non-nuclear weapon states, their decision to do so would exert positive effects on many national non-proliferation policy settings to a degree which would be disproportionate to their costs. Should the nuclear weapon states bite the nuclear inspections bullet and admit the IAEA Inspectorate to all and any of their civilian nuclear facilities, it would be reasonable to expect an immediate dividend in the form of reduced levels of suspicion in all other states on issues such as policy transparency, real non-proliferation intentions, and the end of concerns from states such as Germany, Japan and Canada over the potential commercial export disadvantage suffered by their reactor industries.

Although the breakdown of this discriminatory element of the terms of the NPT was widely expected in the aftermath of the end of the Cold War, early disappointment in the non-nuclear weapon states over the failure to move forward on it has not dissipated. The mooted multilateral Fissile Material Cut-off Treaty, which would have demanded IAEA inspections to verify that the nuclear weapon states were not simply returning surplus weapons materials to new nuclear programs, did not materialize. This was partly due to the US withdrawal in 2004 of support for a verified FMCT treaty, and partly because of the non-appearance of the proposed Trilateral Initiative calling for IAEA verification that the US and Russia were not, in fact, returning "excess" fissile material back to weapons programs.

The issue of comprehensive and mandatory extension of verification inspections into the nuclear weapon states today remains extremely controversial, especially in rapidly industrializing countries such as India and Brazil, in view of the fact that much of the world's nuclear fuel reprocessing and enrichment capacity, as well as most HEU and separated plutonium, is located in the nuclear weapon states. These five states are highly likely to figure prominently in any processing (and reprocessing) of foreign spent nuclear fuel under international fuel bank arrangements and should, in the view of many other states, come under rigorous inspection regimes as a result.[3]

Once again, only time will tell whether, and if so the extent to which, the nuclear weapon states will prove willing to renounce the distinct and well guarded advantages of not having to submit their nuclear activities to the scrutiny of the world. But they ought to do so, if only for the large scale and durable non-proliferation and security benefits available to themselves from so doing. The United States could, and should, now grasp the opportunity to demonstrate to nuclear aspirants such as India, Pakistan, Iran, North Korea and Israel that the advantages of IAEA safeguards truly do outweigh the eventual costs.

The Security Effects of a Civil Nuclear Renaissance

The extent of renewed interest in nuclear power generation which is presently under way globally is evident from the following data, assembled by the World Nuclear Association. The WNA points out that the world's fleet of commercial nuclear power generation reactors stood in March 2009 at 436, located in thirty countries and providing around 15 per cent of the world's total electricity. As a comparison, the total output of over 370,000 MW(e) (megawatts electric) is more than three times the generating capacity of France or Germany from all sources.

Some 30 new reactors are currently under construction, while over 90 more are firmly planned, which is equivalent to adding a further 35 per cent to total world nuclear generating capacity. The United States currently relies on nuclear power for around one fifth of its electrical energy needs, which itself is equivalent to nearly one third of global nuclear power generation. On a league table of global electricity generation sources, the lead is held by black and brown coal with 40.8 per cent, followed by gas (20 per cent), hydro (16.4 per cent), nuclear (14.7 per cent), oil (5.8 per cent) and "other," including non-hydro renewable sources (2.3 per cent).[4]

As nuclear power generation begins rapidly to grow, so too does the potential for nuclear proliferation generally, and in particular the security of nuclear fuels. Proliferation concerns extend to both nuclear terrorism and state-based nuclear aspirations, and are sited at both the front end of the nuclear fuel cycle (in the

---

3 Global fissile material report 2007, International Panel on Fissile Materials, at p. 67–73. Available online at: http://www.fissilematerials.org/ipfm/site_down/gfmr07.pdf. Accessed September 19, 2009.

4 Nuclear power in the world today, *World Nuclear Association*, March, 2009. Available online at: http://www.world-nuclear.org/info/inf01.html. Accessed September 19, 2009.

production of nuclear fuel involving uranium enrichment) and at the back end of the cycle (in the management of plutonium in spent fuel).

As far as fuel enrichment is concerned, this now occurs primarily in the United States, Western Europe and Russia, where sufficient capacity exists to meet the projected five fold increase in demand for low enriched uranium (LEU). This leaves open the slim but real possibility of the successful introduction of an international division of the nuclear world between fuel supplier and consumer states although, as previously discussed, such an outcome will more likely than not be frustrated (at least in its early formats) by the sovereign intentions of states such as India and Iran.

At the back end of the fuel cycle, the consequences of strong growth in the nuclear power generation industry centers around the 75 tons of plutonium contained in the 10,000 tons of spent fuel discharged each year from nuclear reactors. As noted in earlier chapters, this material is managed in three ways:

- reprocessing to recover the plutonium for use in new mixed oxide (MOX) fuel
- indefinite storage with a view to possible future use in fast breeder or burner reactors
- interim storage awaiting geological disposal in a repository, or later reprocessing.[5]

The reprocessing option is growing progressively less popular for nuclear generating countries which previously sent their spent fuel to major reprocessing states such as the United Kingdom, Japan, France and Russia. At least twelve of these have now switched to interim storage (using the dry cask storage method) which they apply to spent fuel no more than twenty years after its discharge on economic cost grounds, but with implications for the security of spent fuel stored in this way.

Any return to the use of reprocessing of spent fuels which virtually eliminate the possibility of fissile materials being obtained by nuclear terrorists may be many years away. On the other hand, an open, or "once through" nuclear fuel cycle, in which there is no bomb-capable material in the fresh reactor fuel, and the plutonium in the spent fuel is irretrievably combined with fiercely radioactive fission products, may well be the most proliferation resistant solution of all.

At the end of the day, though, a growing interest in nuclear power generation necessarily involves the emergence of newly nuclear interested states, some of which may yet be tempted towards the possibilities of fissile material diversion, with all the nuclear security implications which accompany such plans. As the 2007 Report of the International Panel on Fissile Material highlights, if a large-scale expansion of nuclear power were accompanied by a parallel shift towards reprocessing and plutonium recycling, annual flows of separated plutonium could

5 Ibid. At Note 3, pp. 88–91.

reach levels of a thousand tons per annum. This quantum of plutonium is enough for some 100,000 nuclear weapons, even a tiny fraction of which could, in the wrong hands, spell catastrophe in a nuclear terrorist attack.

Nuclear Energy and Climate Change

Nuclear energy production, and the phenomenon of global atmospheric warming leading to climate change, are extremely controversial realities for the new century. The fact that they are also closely linked provides the most promising and realistic opportunity for nuclear power to contribute maximally to reducing carbon emissions from large scale, base load static energy production.

The issue most relevant to this discussion of the nuclear fuel security effects of a nuclear renaissance, however, is the capacity for nuclear aspirant states to accept the internationalization of nuclear fuel supply services as a necessary guarantee of both their own energy security, and the contribution that a nuclear power industry with fuel guarantees would make towards reducing global atmospheric carbon emissions. Those states which are willing to support a viable international fuel bank proposal of the kind outlined recently by the retiring Director General of the International Atomic Energy Agency, Dr Mohamed ElBaradei, and discussed in Chapter 8, will enjoy significant advantages over resistant states.

First, they will take direct steps to reinforce their energy security through diversification of available energy sources into economically viable and fuel source-guaranteed nuclear electricity generation. Second, they will thereby contribute to the global retreat away from the world's high carbon emission energy economy and towards environmentally and economically sustainable low total emissions alternatives. Third, they will also have contributed immensely to global measures to rein in, and ultimately to deny to nuclear terrorists and weapons proliferating states, the means and opportunity to illicitly obtain fissile materials from inadequately secured national or sovereign sources.

Each one of these three results is a worthwhile outcome for such states, as it is for the international society of states as a whole. Together, they form an agenda whose beneficent potential may never return, as the potential for a nuclear terrorist nightmare grows ever closer. It is therefore extremely important that those states which are able swiftly to judge for themselves the net benefits available from complete nuclear fuel security, can *also* be persuaded to come together in functional alliances, formed within the existing international energy and nuclear non-proliferation security policy frameworks, to bring about this enormous policy change through cooperative international action. So long as sufficient support for this strategic move exists within the five NPT nuclear weapon states, and among other important nuclear states (such as Japan, India and Germany) there is no insurmountable barrier, whether political, legal, economic, environmental or otherwise, to its success.

The question for resolution is, therefore, whether sufficient, and sufficiently strong, evidence exists to support widespread acceptance of the energy-economy-environment nexus in respect of a global nuclear fuel bank initiative which irrevocably divides the nuclear energy world between suppliers and consumers of nuclear fuel. This is a judgment which, in the final analysis, only the peoples and leaders of the nation states, taken singly but acting in alliance, can make. However, much high quality scientific and social scientific research is available to allow informed decisions to be made on each of the tripartite limbs of the question. And, as the final paragraph of the Government of the United Kingdom's Stern Review of the Economics of Climate Change (one such extremely authoritative source) notes:

> Above all, reducing the risks of climate change requires collective action. It requires co-operation between countries, through international frameworks that support the achievement of shared goals. It requires a partnership between the public and private sector, working with civil society and with individuals. It is still possible to avoid the worst impacts of climate change, through strong collective action starting from now.[6]

As this analysis has argued strongly, the only viable long-term solution to the dual security problems of nuclear terrorism and state-based nuclear weapons proliferation is one based on the security of commercial nuclear fuels and other fissile materials. This extends to its physical security as it is being prepared for use (for example, in the fabrication of fuel assemblies), when in use at nuclear power generating plants, in storage at nuclear generating facilities, and in transit between facilities. It also includes the transparent and verifiably proliferation-resistant management of fissile material to avoid any possibility for theft or diversion to military programs. This goal will only be reached in the context of closely integrated international architectures and system in which each element of the management of fissile material is tracked and verified safe using the combined security resources of the nuclear world and beyond. An immensely important element of that goal is the construction of workable international fuel bank services, built on trust and (just as important) verified through more comprehensive and rigorous inspection activities than ever before. Such inspections must be undertaken by realistically resourced international bureaucracies, the most important being, of course, the IAEA.

---

6 *Stern review of the economics of climate change*, Part VI, International Collective Action, clause 27.5: Conclusions. Available online at: http://webarchive.nationalarchives.gov.uk/+/http://www.hm-treasury.gov.uk/media/F/8/Chapter_27_Conclusions.pdf. Accessed September 20, 2009.

A Final Observation

On the elements of trust between states which accompany all unilateral and coordinated nuclear activities, there can little doubt for those of goodwill and reason that nuclear proliferation (if allowed to grow uncontrolled) most assuredly breeds security dilemmas of the kind which have afflicted North East Asia for decades. But the only imagined solution in a strategic security system such as this may well amount to further nuclear proliferation. In the world of sovereign states, existential nuclear threats, whether real or reflexively constructed, cannot be ignored by any national government worthy of the name. National security is, first, last and always, their primary duty and *sine qua non*. This component of nuclear [in]security produced the Cold War, and its re-emergence in regional settings must be snuffed out as it appears.

The final chapter of this analysis will discuss and critique the difficulties inherent in durable and traditionally understood nostrums which are a feature of all mature intellectual and pragmatic domains. On this occasion they appear in the form, for example, of the suite of well-established, even entrenched, targets of nuclear fuel security analysis and policy action which have been the subject of much of this analysis. They also appear as acceptable hierarchies of importance within nuclear threat scenarios which contribute, through settled security policy processes, to the allocation of scarce resources in response. Judgments which involve this triage process materially influence the making of national policy, where limited resources are confronted with unlimited need. In the case of the homeland security policy of the United States, the danger of failing to determine the most effective allocation of resources to identified risks could have the most horrendous of security consequences.

While not seeking in an unwarranted fashion to minimize the suite of nuclear risks, Chapter 10 considers in a balanced way some of the apparent assumptions underpinning the risk assessments of nuclear security policy planners, and the bases of those assumptions. One important part of such considerations is the fallibility of systems of intelligence gathering and analysis. This was revealed by the intelligence failures which accompanied the 2003 invasion of Iraq by the Coalition of the Willing. In that case, it is reasonable to question the quality of the intelligence data used and its analysis, which together erroneously identified a mature Iraqi nuclear weapons program which did not exist.

The final chapter will, however, take a cautiously optimistic view of the chances for success of emerging nuclear non-proliferation strategies, and especially of the move to take far closer control over the provision of fuel for nuclear power generation. The central implication here is the final removal of any viable or legitimate reason for states with bona fide aspirations to a nuclear power industry to indulge in the national enrichment of uranium, or the separation of plutonium from spent fuel discharged from their production reactors.

Chapter 10
Conclusions: Global Threats and Opportunities

Introduction

The purpose of the final chapter is to add notes of both caution and optimism to earlier analysis and conclusions presented within this extended discussion. The plethora of recent and continuing US and internationally-led policy changes, and the pattern of emerging initiatives in the field of nuclear fuel security reflected in these pages, are symptomatic of the rapid pace of development in the field. In general terms, as threat levels are continuously reviewed and reassessed in the light of evolving intelligence estimates and analysis, so policy responses are refashioned within the national security assessment machinery of government. In the field of securing the nation against the effects of nuclear weapons proliferation and possible acts of nuclear terrorism, the level of delay which is generally acceptable within this process must often be reduced to that judged to be the absolute minimum necessary for secure, effective and accountable policy-building and implementation.

In a policy assessment process such as this, which is defined by the unassailable importance of its subject matter, urgency, haste and lack of time to test and monitor policy modifications may mean that misjudgments and faulty assessments, as well as inefficient implementation of often partially ineffectual policy settings, are practically unavoidable. This remains true even in circumstances in which mooted change is clearly one part of a broader policy trend extending over the medium term. As time presses, and as political focus is inevitably readjusted (perhaps as a result of a change of political leadership) so human fallibility inevitably translates, in some degree, into policy fallibility.

The chapter incorporates a brief examination of the pressure points of policy change and implementation in the case of the rapidly evolving nuclear fuel security policy field. It does so in the hope that it will draw more attention to the possibility, however unwelcome, that *this* part of America's broad nuclear security and homeland security policy complex may not always be ideally, or even acceptably, effective. These pressure points or policy vulnerabilities are reflected in the suite of nuclear proliferation initiatives of the Obama Administration, announced by President Obama at the 64th Session of the United Nations General Assembly in September, 2009, and incorporated in the US-sponsored United Nations Security Council Resolution 1887 (2009) of September 24, 2009. This welcome development, discussed below, is a serendipitous reminder of the growing urgency with which the international community of states now views effective global nuclear security, as noted many times in this analysis. Certainly, much vital policy work remains to be accomplished, but these initiatives, and the policy renewal they represent, also serve as a reminder that the US can (and must) become a more engaged and flexible leader of change in the evolving processes of global nuclear security policy which will lead to a safer world.

Policy Change and Response

A recent example of political and international security dynamics resulting in adjustments in strategic security policy trends is relevant to a cautious approach to nuclear fuel security policy change. The political context is the advent of the new Obama Administration in Washington. The strategic security context is the gradual adjustment of focus and resources away from homeland security policy settings incorporating long-range missile defense systems, and towards the shorter-range missile defense of allies and US forces in overseas theaters.

More specifically, it concerns President Obama's decision, announced in September, 2009, to abandon US deployment of an advanced radar and associated missile interceptors, to be fielded in Poland and the Czech Republic against long-range Iranian missile attack, and originally planned for completion by 2017. US Defense Secretary Robert M. Gates noted at the time (as he was bound to do) that homeland defense remained a high priority, with long-range interceptors still planned for installation in Alaska and California against a possible North Korean missile attack.

Critics of this alteration in US strategic security planning pointed out that its implementation carried the risk of serious imbalance in the allocation of resources for countering the full extent of possible missile threats to the United States, including long-range missile attack. Robert G. Joseph, former Undersecretary of State for Arms Control and International Security during the final Bush Administration, who was an architect of the Proliferation Security Initiative and remains a strong advocate for missile defense, has asserted that this strategic move within the European theater "weakens our capability against long-range threats." He has also stated his belief that any decision not to protect the United States from blackmail and intimidation by foreign leaders armed with intercontinental ballistic missiles would amount to a reliance on "dangerously outdated" theories of warfare.[1]

1 Smith, R. Jeffery, Missile defense's shelving reflected Military's concerns, *The Washington Post*, Monday, September 21, 2009. Available online at: http://www.washingtonpost.com/wp-dyn/content/article/2009/09/20/AR2009092002532_2.html?sid=ST2009092002872. Accessed September 21, 2009. It is worth noting that the decision of the Obama Administration to abandon European-based missile defense systems resulted, for example, in the immediate thawing of the US/Russia relationship. One tangible effect

The central lesson to be taken from all of this, at least in the context of nuclear fuel security, is that perceptions of what a security challenge really amounts to are not immutable, nor are they unlikely to change over time in material ways. Rather, threat perceptions held by those individual and collective agents charged with constructing and implementing appropriate policy responses are the result of multiple influences and inputs. These may include personal or corporate belief and conviction, individual experience and preference, organizational culture, statutory duty, national fealty and many more. Whatever their source, however, such determinants of agency and identity result in a world view which materially conditions the range of responses acceptable to a change agent in reaction to partially subjective perceptions of threat. To return briefly to the earlier 2003 Iraq nuclear intelligence example, such a significant failure of policy process and its intelligence drivers is always a possibility when the presence of complexly conditioned perception of risk is linked to political or military decision authority.

Having identified all elements of a danger in specific terms, in this case the ultimate threat of nuclear terrorism, the next step is, of course, to determine the actions necessary in response. It is at this point that the range of possible policy responses opens up in many directions, usually emerging as policy action alternatives, although often tending thereby to set up internally inefficient and overlapping policy competition. The difficulties inherent in coordinating several policy responses with similar goals, but relevant in varying time frames, is only one of the many challenges of the government policy change process. Furthermore, this is, unfortunately, no less likely in the international policy sense than it is of domestic policy change and evolution. In both fields, totally effective coordination of policy aims, and their implementation over time, is almost certainly impossible, although the capacity of oversight authority to monitor policy effectiveness and modify action as appropriate is certainly capable of moderating the worst inefficiencies present in resource usage.

With these cautionary general reflections on the limitations of policy action, and with a lead from a rejuvenated US nuclear security environment, it is possible to more clearly identify where the solutions to US nuclear fuel security policy weaknesses may lie.

US Nuclear Fuel Security Policy Vulnerabilities

In the final analysis, ideally effective nuclear fuel security policy, like all other government policy fields, continuously confronts the economic imperative. Only so many resources are available, at the conclusion of the annual federal budget process, for allocation to each area of government policy and action, since government resources are, of course, finite. Inevitably, some policy fields will be

has been Russia's President Medvedev's support for UN Security Council Resolution 1887 on steps to further curb nuclear proliferation (to be discussed later in this chapter).

under-resourced or otherwise starved of funds, while others (for reasons of the kind noted above, and including political ones) will be more generously funded.

To extend this line of argument further, preferences in resource allocation are more or less likely to be expressed by financial support for policy initiatives according to the extent to which clearly demonstrated and verifiably achievable goals are available within specific and achievable time frames, and are based on coherent and easily understood resource allocation methodologies. Furthermore, the extent to which this level of funding coherence is ever achieved will normally be a function of the success of a competing range of less relevant, and often politically driven, policy alternatives.

United States nuclear fuel security policy in pursuit of protection against the risks of nuclear terrorism is, unfortunately, not in the first – essentially theoretical – category. Instead, it presents a highly complex policy challenge whose goal is the attainment of a negative outcome. Through the *absence* of the threat of, or an actual, nuclear attack, it strives to reach a level of security which fully protects the United States and its people against all credible risk of terrorist or other attack upon the American homeland using fissile materials and nuclear weapons.

The reality, however, is that it is generally extremely difficult to demonstrate to skeptical financial gatekeepers in Congress a convincing policy strategy leading to the achievement a negative result, and no less so within the specific policy parameters of nuclear fuel security, however serious the consequences of failure. From this difficulty comes the essentially impossible feat of definitively identifying the suite of policies which together created the absence of previous risk, and thus success.

Added to this inherent difficulty, in this case, is the sheer complexity of the policy task, which continues to grow, and which is reflected throughout this book. The combined effect of these impediments on coherent US nuclear security policy development, both in the domestic and global spheres, is that it tends to follow the path, and fate, of many other complex policy fields: relative policy incoherence. This is so despite the appearance in 2003 of the Department of Homeland Security (DHS) through the merger of twenty-two individual government agencies, and charged with the protection of the United States within, at, and outside its borders. With the aim, among its many other responsibilities, of preparing for, preventing and responding to acts of terrorism within United States territory, the DHS is confronted with the reality that it constitutes only one part of the machinery of government with responsibilities in respect of nuclear terrorism.

The plain fact is that there exists *no* method within the United States federal government of accurately calculating the most effective allocation of resources to repel the nuclear terrorist threat among myriad claims from many sources; no feasible way to compare marginal benefits in terms of the reduction of threat levels to be expected from added spending in one policy area, rather than another. Any such aspiration is defeated by the extreme uncertainties surrounding the nuclear terrorist risk profile. Moreover, a serious and open attempt at such a calculus would be more likely than not to create even greater uncertainties as to the effectiveness of preferred policy initiatives by fueling doubts, at least among disappointed advocates, about the opportunity cost of failing to adequately fund, or even abandon, other programs. As in no field other than state-based threats founded on nuclear weapons proliferation, the price of nuclear security policy failure may well be unbearable – for both the United States and the world.

Given the existential nature of the threat, far too little has been attempted within the relevant agencies of the US government to address this issue in even a partial, much less whole-of-government way. Should it ever be undertaken in a systematic way, progress would require a fully comprehensive analysis across departments, agencies and programs to identify areas in urgent need of additional funds, and their available sources from beyond specific agency budgets. There is no doubt that a root-and-branch exercise of this kind would meet with resistance ranging from the intense to the specious, much of it generated by internal government turf wars and mutual animosity. Even give the absence of the dead hand of bureaucracy, the re-direction of funding driven solely by the criterion of expressed need will not, of course, solve the problem of mis-direction of funding emphasis, but may at least alleviate the influence of some non-relevant funding criteria.

On the broader government sector scale, this kind of cross-agency and cross-program strategic approach to coordinated policy responses across government is essentially alien to the traditional processes of funding allocation within a highly politicized Washington budget process. Often mired within Congress in the complexities of local politics, the personal and cultural dynamics of Congressional Committees, bureaucratic combat for funding, and parallel but uncoordinated Office of Management and Budget (OMB) oversight, the probable fate of a coherent and coordinated nuclear fuel security policy is easy to comprehend on the funding level. It is less easy to reconcile with the ultimate cost to the American nation of getting nuclear security wrong.[2]

The Ultimate Price of Failure

Chapter 1 drew readers' attention to the major physical attributes of nuclear explosions using highly enriched uranium or weapons-grade plutonium, divided between fireball, flash, ground shock and electro-magnetic pulse. Its generalized description of the four main effects was intended only as a simple introduction to the unique technical and physical aspects of the military use of explosive nuclear fission. As such, it gave little indication of the combined effects of a nuclear detonation on people and physical objects.

It seems clear that, if this book's discussion of nuclear fuel security and its impact on US homeland security is to be fully understood in terms of the *effects* of

2 Bunn, Matthew. *Securing the bomb 2008*. Cambridge, Mass. and Washington, DC: Project on Managing the Atom, Belfer Center for Science and International Affairs, Harvard Kennedy School and Nuclear Threat Initiative, November 18, 2008. At pp. 124–25.

nuclear explosions detonated, for example, over a large city, it should include that information in neutral, unadorned form. Those who help to put into practice the panoply of nuclear security-related policy decisions, at home and abroad, should do so in the full knowledge not only of what they are seeking to protect and secure, but also of the possible consequences for America's people should those policies fail.

With that in mind, what follows is intended for use as an aid to memory, to be consulted when the routine language of "nuclear threats" and "nuclear terrorism," so often deployed unthinkingly in the news cycles of current affairs media and elsewhere, begins to lose all but its iconic connotations. The bulk of this information is taken from The Nuclear Weapons Archive: A Guide to Nuclear Weapons and the Federation of American Scientists.[3]

The Nature of a Nuclear Explosion

Nuclear detonations are the most devastating of the weapons of mass destruction. The images recorded at Hiroshima and Nagasaki, and the international furore over the accidental but enormous radiation release in 1986 from the Chernobyl nuclear power plant in Ukraine, confirms this. The level of contamination from the Chernobyl accident was significantly larger than would have been expected from a nuclear detonation of about 20 kt at ground level, but was comparable in extent to what might result from a dozen or so weapons of nominal yield exploded at altitudes intended to maximize blast damage.

A nuclear detonation creates a devastating environment that includes blast, thermal pulse, neutron radiation, x- and gamma-ray radiation, an electromagnetic pulse (EMP), and ionization of the upper atmosphere. Depending on the environment in which the device is detonated, blast effects are experienced as ground shock, water shock, cratering, and large amounts of dust and radioactive fallout. The energy of a nuclear explosion is transferred to the surrounding medium in three distinct forms: blast; thermal radiation; and nuclear radiation. The distribution of energy among these three forms will depend on the yield of the weapon, the location of the burst, and the characteristics of the environment. For a low altitude atmospheric detonation of a moderately sized weapon in the kiloton range, the energy is distributed roughly as follows:

- Blast – 50 per cent
- thermal radiation – 35 per cent, made up of a wide range of the electro-

3 Sublette, C., *Section 5.0: Effects of nuclear explosions, Nuclear weapons frequently asked questions,* The nuclear weapons archive: A guide to nuclear weapons, Version 2.14, May 15, 1997. Available online at: http://nuclearweaponarchive.org/. Accessed September 28, 2009. *Nuclear weapon effects, Special weapons primer, Weapons of mass destruction*, Federation of American Scientists, October 21, 1998. Available online at: http://www.fas.org/nuke/intro/nuke/effects.htm. Accessed September 29, 2009.

magnetic spectrum, including infrared, visible and ultraviolet light, and some soft x-rays emitted at the time of the explosion

- nuclear radiation – 15 per cent, including 5 per cent as initial ionizing radiation, consisting mainly of neutrons and gamma rays emitted within the first minute after detonation, and 10 per cent as residual nuclear radiation, to be precipitated later as fallout.

Considerable variation from this distribution will occur with changes in the yield or location of the detonation.

Because of the tremendous amounts of energy liberated per unit mass in a nuclear detonation, temperatures of several tens of millions of degrees centigrade develop in the immediate area of the detonation. This contrasts with the few thousand degrees of a conventional explosion. At these very high temperatures the non-fissioning parts of the nuclear weapon are immediately vaporized.

The fission process releases energy not as kinetic energy but in the form of large amounts of electromagnetic radiation. In an atmospheric detonation, this electromagnetic radiation, consisting chiefly of soft x-rays, is absorbed within a few meters of the point of detonation by the surrounding atmosphere, heating it to extremely high temperatures and forming a brilliantly hot sphere of air and gaseous weapon residues, known as the fireball.

Immediately upon its formation, the fireball begins to grow rapidly and rise like a hot air balloon. Within a millisecond after detonation, the diameter of the fireball from a 1 megaton (Mt) air burst is around 150 meters. This increases to a maximum of some 2,200 meters within 10 seconds, at which time the fireball is also rising at the rate of 100 m/sec. The initial rapid expansion of the fireball severely compresses the surrounding atmosphere, producing a powerful blast wave.

As it expands toward its maximum diameter, the fireball cools, and after about one minute, its temperature has decreased to such an extent that it no longer emits significant amounts of thermal radiation. The combination of the upward movement and the cooling of the fireball gives rise to the formation of the characteristic mushroom-shaped cloud. As the fireball cools, the vaporized materials in it condense to form a cloud of solid particles.

Following an air burst, condensed droplets of water give it a typically white, cloud-like appearance. In the case of a surface burst, this cloud will also contain large quantities of dirt and other debris which were vaporized when the fireball touched the earth's surface, or were sucked up by the strong updrafts afterwards, giving the cloud a dirty brown appearance. The dirt and debris become contaminated with the radioisotopes generated by the explosion or activated by neutron radiation and fall to earth as fallout. The megaton class weapons that were developed in the US and Soviet Union during the nineteen-fifties and sixties have been largely retired, being replaced with much smaller yield warheads. The yield of a modern strategic warhead is, with few exceptions, now typically in the range of 200–750 kt. Recent work with sophisticated climate models has shown that this reduction in yield results in a much larger proportion of the fallout being deposited in the lower atmosphere, and a much faster and more intense deposition of fallout than had been assumed in studies made during the nineteen-sixties and seventies. The reduction in aggregate strategic arsenal yield that occurred when high yield weapons were retired in favor of more numerous lower yield weapons has actually increased the fallout risk.

The relative effects of blast, heat, and nuclear radiation will largely be determined by the altitude at which the weapon is detonated. Nuclear explosions are generally classified as air bursts, surface bursts, subsurface bursts, or high altitude bursts.

Air bursts

An air burst is an explosion in which a weapon is detonated in air at an altitude below 30 km but at sufficient height that the fireball does not contact the surface of the earth. After such a burst, blast may cause considerable damage and injury. The altitude of an air burst can be varied to obtain maximum blast effects, maximum thermal effects, desired radiation effects, or a balanced combination of these effects.

Burns to exposed skin may be produced over many square kilometers and eye injuries over a still larger area. Initial nuclear radiation will be a significant hazard with smaller weapons, but the fallout hazard can be ignored, as there is essentially no local fallout from an air burst. The fission products are generally dispersed over a large area of the globe unless there is local rainfall resulting in localized fallout. In the vicinity of ground zero, there may be a small area of neutron-induced activity which could be hazardous to people required to pass through the area.

Surface burst

A surface burst is an explosion in which a weapon is detonated on, or slightly above, the surface of the earth so that the fireball actually touches the land or water surface beneath it. Under these conditions, the area affected by blast, thermal radiation, and initial nuclear radiation will be less extensive than for an air burst of similar yield, except in the region of ground zero where destruction is concentrated. In contrast with air bursts, local fallout can be a hazard over a much larger downwind area than that which is affected by blast and thermal radiation.

Sub-surface burst

A subsurface burst is an explosion in which the point of the detonation is beneath the surface of land or water. Cratering will generally result from an underground burst, just as for a surface burst. If the burst does not penetrate the surface, the only other hazard will be from ground or water shock. If the burst is shallow enough to penetrate the surface, blast, thermal, and initial nuclear radiation effects will be present, but will be less than for a surface burst of comparable yield. Local fallout will be very heavy if such penetration occurs.

High altitude burst

A high altitude burst is one in which the weapon is exploded at such an altitude (above 30 km) that initial soft x-rays generated by the detonation dissipate energy as heat in a much larger volume of air molecules. In this case, the fireball is much larger and expands much more rapidly. The ionizing radiation from a high altitude burst can travel for hundreds of miles before being absorbed by the atmosphere. Significant ionization of the ionosphere can occur. Severe disruption in communications may also occur following high altitude bursts through the generation of an intense electromagnetic pulse (EMP) which can significantly degrade the performance of sophisticated electronic equipment, or destroy it. Similar effects can be expected in any industrial or commercial equipment dependent on digital information processing for its operation, including transportation equipment.

Although some nuclear weapons effects such as blast and cratering have analogs in the effects of conventional weapons, many, such as the EMP effect, are unique to nuclear weapons use. In addition, blast and other conventional weapons effects are likely to be much more powerful in the nuclear case.

Although thermal radiation, EMP, and ionizing radiation from a nuclear blast are all damage producing, at yields below about one megaton the blast and shock effects produced by a nuclear weapon are the predominant means of damaging a target. For some targets, such as underground bunkers and missile silos, blast and shock are virtually the only effective destructive mechanisms.

The intensity of thermal radiation decreases only as the inverse square of the distance from a nuclear detonation, while blast, shock, and prompt ionizing radiation effects decrease more rapidly. Thus, high-yield weapons are primarily incendiary weapons, able to start fires and do other thermal damage at distances well beyond the radius at which they can topple buildings or overturn armored vehicles.

The electromagnetic pulse generated by the detonation of a single nuclear weapon at high altitudes can also be a threat to military systems located as much as a thousand miles away, and can disable communications systems and even power grids at enormous distances from the burst. The use of a high-yield thermo-nuclear device (of 1 Mt or more) exploded in the upper atmosphere may well be the most effective strategy for any state wishing to cause the maximum possible military, social and economic damage to an adversary state.

Immediate and Delayed Destructive Effects

Nuclear explosions produce both immediate and delayed destructive effects. Immediate effects (blast, thermal radiation, prompt ionizing radiation) are produced and cause significant destruction within seconds or minutes of a nuclear detonation. The delayed effects (radioactive fallout and other possible environmental effects) inflict damage over an extended period ranging from hours to centuries, and can cause adverse effects in locations distant from the site of the detonation.

Overview of Immediate Effects

The three categories of immediate effects are the blast, thermal radiation (heat), and prompt ionizing or nuclear radiation discussed earlier. Their relative importance varies with the yield of the bomb. At low yields, all three can be significant sources of injury and, at a very small explosive yield of about 2.5 kt, the three effects are roughly equal. All are capable of inflicting fatal injuries at a range of 1 km.

In the Hiroshima attack (with a yield of approximately 15 kt) casualties (including fatalities) were seen from all three causes. Burns, including those caused by the ensuing fire storm, were the most prevalent serious injury (two-thirds of those who died on the first day succumbed to burns injuries), and occurred at the greatest range. Blast and burn injuries were both found in some 65 per cent of all survivors. People close enough to suffer significant radiation illness were well inside the lethal effects radius for blast and flash burns and, as a result, only 30 per cent of injured survivors showed radiation illness. Many were sheltered from burns and blast, and thus escaped their main effects. Even so, most victims with radiation illness also had blast injuries or burns.

With yields in the range of hundreds of kilotons or greater, immediate radiation injury becomes insignificant. Dangerous radiation levels only exist so close to the explosion that surviving the blast is impossible. On the other hand, fatal burns can be inflicted well beyond the range of substantial blast damage. A 20 megaton bomb, for example, can cause potentially fatal third degree burns at a range of 40 km, where the blast can do little more than break windows and cause superficial cuts.

It should be noted that the fission bombing of Hiroshima and Nagasaki caused fatality rates which were one to two orders of magnitude higher than the rates seen in conventional fire raids on other Japanese cities made during the closing months of the Second World War. Eventually, around 200,000 fatalities, which equates to about one-quarter of all Japanese bombing deaths, occurred in these two cities, with a combined population of less than 500,000. This was due to the fact that the bombs inflicted damage on people and buildings virtually instantaneously and without warning, and did so with the combined effects of flash, blast, and radiation. Widespread fatal injuries were thus inflicted instantly, while many more people were incapacitated and thus unable to escape the rapidly developing fires in the suddenly ruined cities.

A rule of thumb for estimating the short-term fatalities from all causes in a nuclear attack is to count everyone within the five pounds per square inch (psi) blast overpressure contour around the explosion's center as a fatality. In reality, substantial numbers of people inside the contour will survive, while substantial numbers outside the contour will die, the assumption being that these two groups will be of roughly equal size. This calculus, however, ignores any possible fallout effects.

Overview of Delayed Effects

The main delayed effect is the creation of huge amounts of radioactive material with long lifetimes, ranging from days to millennia. The primary source of these fission products is the debris left from the fission chain reaction that creates the bomb's detonation. A potentially significant secondary source of radiation is neutron capture by non-radioactive isotopes, both within the bomb and in the external environment.

When atoms fission, they can split in some forty different ways, producing a mix of about eighty different isotopes. These isotopes vary widely in stability, some being completely stable while others undergo radioactive decay with half-lives of fractions of a second. The decaying isotopes may themselves form stable or unstable "daughter" isotopes. The mixture of dangerously radioactive fission products thus quickly becomes even more complex, some 300 different isotopes of thirty-six elements having been identified.

Short-lived isotopes release their decay energy rapidly, creating intense radiation fields that also decline quickly. Long-lived isotopes release energy slowly, over long periods of time, creating radiation that is much less intense but more persistent. Many fission products of a nuclear fission explosion initially have very high levels of radiation that declines quickly, although as the intensity of radiation drops, so too does the rate of decline. A useful rule-of-thumb is the "rule of sevens," which states that for every seven-fold increase in the passage of time following a fission detonation (starting at or after one hour), the radiation intensity decreases by a factor of ten.

Radioactive fission products are most hazardous when they settle to the ground as "fallout". The rate at which fallout settles depends closely on the altitude at which the explosion occurs, and to a lesser extent on the size of the explosion. If the explosion is a true air-burst (in which the fireball does not touch the ground), when the vaporized radioactive products cool sufficiently to condense and solidify, they form microscopic particles. These particles are mostly lifted high into the atmosphere by the rising fireball, although significant amounts are deposited in the lower atmosphere by mixing that occurs due to convective circulation within the fireball.

The larger the explosion, the higher and faster the fallout is lifted, and the smaller the proportion that is deposited in the lower atmosphere. For explosions with yields of 100 kt or less, the fireball does not rise above the troposphere, where natural precipitation occurs. All of this fallout will thus be brought to the ground by normal weather processes within months of the explosion, at most, and usually much faster. In the megaton range, the fireball rises so high that it enters the stratosphere, where the air is dry, and no weather processes exist to bring fallout down quickly. In this case, small fallout particles will descend over a period of months or years. Such long-delayed fallout has lost most of its radiation by the time it reaches the ground, and will be distributed on a global scale. As yields increase above 100 kt, progressively more and more of the total fallout is injected into the stratosphere.

As noted earlier, an explosion which is low enough in the atmosphere for the fireball to touch the ground sucks large amounts of earth into the rising fireball. The earth usually does not vaporize, and if it does, there is so much of it that it forms large particles. The radioactive isotopes are deposited on soil particles, which can fall quickly to earth in a time span measured from minutes to days, creating downwind contamination both nearby and thousands of kilometers away. The most intense radiation is created by nearby fallout, because it is more densely deposited, and because its short-lived isotope content has not yet decayed. Weather conditions will, of course, affect fallout behavior considerably. In particular, rainfall events can create very intense localized fallout concentrations. Both external exposure to penetrating radiation and internal exposure through the ingestion of radioactive material pose serious health risks.

Finally, explosions close to the ground that do not touch it can still generate substantial hazards immediately below the burst point by neutron-activation. Neutrons absorbed by the soil can generate considerable radiation for several hours.

The many severe and fatal injury effects of exposure to a nuclear explosion, such as thermal, eye and blast injury, and the many effects of exposure to ionizing radiation, need not be elaborated here. Instead, the reader is invited to contemplate the full horror of the detonation of a nuclear explosive device in or above a major center of population. The simultaneous destruction of lives and of a way of life, of homes and futures, of both cities and complacency; these are the real faces of the ultimate price of nuclear security failure.

The Road to Opportunity

Fortunately, the picture is not universally bleak. The new Obama Administration's work towards a more active approach to US engagement with global nuclear security challenges, in contrast to the more confrontational policy stance of its predecessor, is a clear and distinctive turn in the nuclear non-proliferation and fuel security road. It provides concrete grounds for optimism about the possibility of substantial progress in cooperative and coordinated action to curb the spread of nuclear weapons, their technologies and fissile materials. After eight years of relative policy stagnation and drift, marked by the skepticism of President George W. Bush and his national security advisors about the value of international arms control agreements, it appears that US nuclear security policy is on the move.

Signs of change are now multiplying. During the 64th Session of the United Nations General Assembly, held in New York during September, 2009, the UN Security Council held only the fifth heads-of-state summit meeting since its institution, chaired by President Obama. Its centerpiece was US-drafted and sponsored Resolution 1887, which affirmed and echoed many of the elements of Obama's program of nuclear policy renewal whose ultimate aim is to rid the world of nuclear arms. There is little doubt that President Obama's presence was intended as a direct signal of the importance his administration attaches to the issue of nuclear security.

Although Obama importantly noted in his address to the Council that "International law is not an empty promise," Resolution 1887 nevertheless accommodates the sensitivities of Russia and China by not referring explicitly to Iran or North Korea in its text as countries of concern in respect of nuclear proliferation and the risk of global nuclear terrorism. Such are the niceties of nuclear diplomacy, at least between the five permanent member states of the Security Council.

In essence, the president's speech to the UN Security Council in support of Resolution 1887 was a reiteration of the points he made in his Prague speech of April, 2009 and noted in Chapter 7. There was, however, an added note of urgency in his rhetoric as he characterized the next twelve months as "absolutely critical" in setting a new world agenda to reverse the spread of nuclear weapons and make progress towards multilateral disarmament. In this context, President Obama restated his commitment to a Global Summit Meeting on Nuclear Security, to be sponsored by the United States and held in New York in March, 2010. Its goals would be to negotiate on securing all vulnerable nuclear materials worldwide within four years, disrupt black markets in nuclear materials and technologies, find ways to enhance the detection and interception of nuclear materials in transit, and use financial tools to track and disrupt illicit trade in nuclear materials.

In order to emphasize the immense significance of UN Security Council Resolution 1887, the first comprehensive UN-based action on nuclear issues since the mid-1990s, consider the extent and scale of its focus, and the actions it takes. Among its most important initiatives, the non-binding resolution:

- Calls on all nuclear weapon states to ratify the Comprehensive Test Ban Treaty (CTBT), which is still not in force because only thirty-one of the required forty-four nuclear-capable countries, but including France, the Russian Federation and the United Kingdom, have ratified it.
- Calls on all states to refrain from nuclear test explosions, and to ratify the CTBT as soon as possible.
- Calls on the nuclear weapon states to negotiate a multilateral Fissile Material Cut-off Treaty (FMCT) at the Geneva Conference on Disarmament, while joining the nuclear disarmament process now being led by the United States and the Russian Federation.

- Calls for several measures to strengthen the Nuclear Non-proliferation Treaty (NPT) ahead of the treaty's five yearly Review Conference in 2010. These include establishing significant costs for states seeking to renounce the treaty and proceed to manufacture nuclear weapons, as North Korea purported to do in 2003.
- Demands that countries involved in "major challenges to the non-proliferation regime" – a veiled reference to North Korea and Iran – comply fully with their obligations under the NPT, and find early negotiated solutions to their nuclear issues.
- Calls for strict controls over nuclear material to ensure that it does not fall into dangerous hands.
- Encourages all member states of the IAEA to ensure the adequacy of the Agency's resources and authority necessary for the detection of the undeclared use of fissile material and verification of the legal use of declared material.
- Urges nuclear exporting countries to make sales of nuclear technologies subject to the IAEA Additional Protocol full-scope inspection safeguards regime.
- Requires the return to exporting countries of nuclear technologies by states withdrawing from the NPT.
- Encourages the work of the IAEA on a multilateral nuclear fuel cycle, including assurances of supply, and urges the Agency's Board of Governors to agree to measures before it to this end as soon as possible.
- Calls for universal adherence to the Convention on the Physical Protection of Nuclear Materials and its 2005 Amendment and the Convention for the Suppression of Acts of Nuclear Terrorism.
- Resolves to monitor the acquisition of nuclear weapons and the means of their delivery by non-state actors, in particular for terrorist purposes, and as defined in UN Security Council Resolution 1540.[4]

In a move which was not unconnected with UNSC Resolution 1887, the leaders of the United States, the United Kingdom and France revealed on September 25 the existence of a previously secret second underground Iranian uranium enrichment facility, to be added to the well known and similar facility located at Natanz and subject to IAEA inspections. Located near the city of Qum, this second facility had long been suspected to be an enrichment facility as a result of US intelligence surveillance, but has now been confirmed as such. Among its alleged attributes are a prospective uranium gas centrifuge cascade of up to 3,000 units, to be fully

4 Security Council SC4976, *Historic summit of Security Council pledges support for stalled efforts to end nuclear weapons proliferation*, United Nations Organization, News and Media Division, Department of Public Information, September 24, 2009. Available online at: http://www.un.org/News/Press/docs//2009/sc9746.doc.htm. Accessed September 26, 2009.

operational within one year and capable of producing sufficient highly enriched uranium annually for one fission weapon.[5]

As a carefully timed move by Presidents Obama and Sarkozy, and Prime Minister Brown, this revelation, if it accomplished nothing else, has reinforced in singular fashion the growing menace of nuclear proliferation in the hands of regimes, or non-state actors, which are determined to leave no stone unturned in their search for acquisition of nuclear explosives. Although Iran had finally revealed the Qum facility to the IAEA days before its public revelation, President Obama stated that it represented "... a direct challenge to the basic foundation of the non-proliferation regime".[6] Further strong economic and possibly political sanctions against Iran are certain to be urged on the Security Council as a consequence of this event.

The confluence of these two events – the unanimous adoption of Security Council Resolution 1887 and the Iranian uranium enrichment revelation – can be regarded, equally convincingly, as grounds for both pessimism or optimism in respect of the chances of non-proliferation and nuclear fuel success on a global scale; pessimism that large-scale nuclear proliferation based on a disregard for international law appears to be continuing unrestrained, or optimism based on the fact that Iran's (and hopefully North Korea's) illicit nuclear activities can, in fact, be detected and reversed, given sufficient political will on all sides.

Similarly, these events can be equally clearly regarded as symptomatic of either a continuing and unassailable threat of existential proportions, or one in respect of which many opportunities exist for its dismantlement and consignment to the pages of history. Each has its adherents, each its opponents. The suite of United States nuclear policy moves which are explicit in President Obama's speeches in Prague, at the United Nations and elsewhere during 2009, and which find their multilateral expression in Security Council Resolution 1887, make plain the choice before all national and community leaders throughout the world. It is a choice between action and inaction, between pessimism and optimism, and between living under threat or resisting it using the opportunity that are so clearly available. Most importantly of all, it is about making a choice.

A Final Word

This book set out its central aim in the Introduction. That was to offer a comprehensive though concise exploration of the effects on security policy of the dangers and threats inherent in the legitimate production, storage, transportation, safeguarding, export and use of highly enriched uranium and plutonium. The

5 Cooper, H., and Mazzetti, M., Cryptic Iranian Note ignited an urgent nuclear strategy debate, *The New York Times: Middle East*, September 25, 2009. Available online at: http://www.nytimes.com/2009/09/26/world/middleeast/26intel.html?hp. Accessed September 26, 2009.

6 Ibid.

readership it has kept in mind throughout has been that of the informed generalist, anyone who has felt that many discussions about nuclear security such as this tend to be unnecessarily laced with too much technical and scientific jargon, or other professional complexities. With the goal of avoiding such disappointment, the book has attempted to steer clear of this kind of approach in the interests of clarity and ease of understanding.

Nevertheless, and as promised in the Introduction, it has undertaken a wide-ranging discussion of interlinked geopolitical, security and technical issues and challenges for national and global governance in the fields of nuclear fuel security, the proliferation of nuclear weapons materials and technologies, and prospects for countering and defeating nuclear terrorism.

Many and growing dangers and threats afflict the Earth, its natural environment and its people. None is more serious or proximate than the group of nuclear-related threats which continue to develop through the misuse of the incredible energy density of nuclear fission. Whether warrior or peacemaker, no-one, anywhere, who cares about the future can afford to ignore this reality.

Appendix A: Treaty on the Non-proliferation of Nuclear Weapons

The States concluding this Treaty, hereinafter referred to as the "Parties to the Treaty",

Considering the devastation that would be visited upon all mankind by a nuclear war and the consequent need to make every effort to avert the danger of such a war and to take measures to safeguard the security of peoples,

Believing that the proliferation of nuclear weapons would seriously enhance the danger of nuclear war,

In conformity with resolutions of the United Nations General Assembly calling for the conclusion of an agreement on the prevention of wider dissemination of nuclear weapons,

Undertaking to co-operate in facilitating the application of International Atomic Energy Agency safeguards on peaceful nuclear activities,

Expressing their support for research, development and other efforts to further the application, within the framework of the International Atomic Energy Agency safeguards system, of the principle of safeguarding effectively the flow of source and special fissionable materials by use of instruments and other techniques at certain strategic points,

Affirming the principle that the benefits of peaceful applications of nuclear technology, including any technological by-products which may be derived by nuclear-weapon States from the development of nuclear explosive devices, should be available for peaceful purposes to all Parties to the Treaty, whether nuclear-weapon or non-nuclear-weapon States,

Convinced that, in furtherance of this principle, all Parties to the Treaty are entitled to participate in the fullest possible exchange of scientific information for, and to contribute alone or in co-operation with other States to, the further development of the applications of atomic energy for peaceful purposes,

Declaring their intention to achieve at the earliest possible date the cessation of the nuclear arms race and to undertake effective measures in the direction of nuclear disarmament,

Urging the co-operation of all States in the attainment of this objective,

Recalling the determination expressed by the Parties to the 1963 Treaty banning nuclear weapon tests in the atmosphere, in outer space and under water in its Preamble to seek to achieve the discontinuance of all test explosions of nuclear weapons for all time and to continue negotiations to this end,

Desiring to further the easing of international tension and the strengthening of trust between States in order to facilitate the cessation of the manufacture of nuclear weapons, the liquidation of all their existing stockpiles, and the elimination from national arsenals of nuclear weapons and the means of their delivery pursuant to a Treaty on general and complete disarmament under strict and effective international control,

Recalling that, in accordance with the Charter of the United Nations, States must refrain in their international relations from the threat or use of force against the territorial integrity or political independence of any State, or in any other manner inconsistent with the Purposes of the United Nations, and that the establishment and maintenance of international peace and security are to be promoted with the least diversion for armaments of the world's human and economic resources,

Have agreed as follows:

ARTICLE I

Each nuclear-weapon State Party to the Treaty undertakes not to transfer to any recipient whatsoever nuclear weapons or other nuclear explosive devices or control over such weapons or explosive devices directly, or indirectly; and not in any way to assist, encourage, or induce any non-nuclear-weapon State to manufacture or otherwise acquire nuclear weapons or other nuclear explosive devices, or control over such weapons or explosive devices.

ARTICLE II

Each non-nuclear-weapon State Party to the Treaty undertakes not to receive the transfer from any transferor whatsoever of nuclear weapons or other nuclear explosive devices or of control over such weapons or explosive devices directly, or indirectly; not to manufacture or otherwise acquire nuclear weapons or other nuclear explosive devices; and not to seek or receive any assistance in the manufacture of nuclear weapons or other nuclear explosive devices.

ARTICLE III

1. Each Non-nuclear-weapon State Party to the Treaty undertakes to accept safeguards, as set forth in an agreement to be negotiated and concluded with the International Atomic Energy Agency in accordance with the Statute of the International Atomic Energy Agency and the Agency's safeguards system, for the exclusive purpose of verification of the fulfilment of its obligations assumed under this Treaty with a view to preventing diversion of nuclear energy from peaceful uses to nuclear weapons or other nuclear explosive devices. Procedures for the safeguards required by this Article shall be followed with respect to source or special fissionable material whether it is being produced, processed or used in any principal nuclear facility or is outside any such facility. The safeguards required by this Article shall be applied on all source or special fissionable material in all peaceful nuclear activities within the territory of such State, under its jurisdiction, or carried out under its control anywhere.

2. Each State Party to the Treaty undertakes not to provide: *(a)* source or special fissionable material, or *(b)* equipment or material especially designed or prepared for the processing, use or production of special fissionable material, to any non-nuclear-weapon State for peaceful purposes, unless the source or special fissionable material shall be subject to the safeguards required by this Article.

3. The safeguards required by this Article shall be implemented in a manner designed to comply with Article IV of this Treaty, and to avoid hampering the economic or technological development of the Parties or international co-operation in the field of peaceful nuclear activities, including the international exchange of nuclear material and equipment for the processing, use or production of nuclear material for peaceful purposes in accordance with the provisions of this Article and the principle of safeguarding set forth in the Preamble of the Treaty.

4. Non-nuclear-weapon States Party to the Treaty shall conclude agreements with the International Atomic Energy Agency to meet the requirements of this Article either individually or together with other States in accordance with the Statute of the International Atomic Energy Agency. Negotiation of such agreements shall commence within 180 days from the original entry into force of this Treaty. For States depositing their instruments of ratification or accession after the 180-day period, negotiation of such agreements shall commence not later than the date of such deposit. Such agreements shall enter into force not later than eighteen months after the date of initiation of negotiations.

ARTICLE IV

1. Nothing in this Treaty shall be interpreted as affecting the inalienable right of all the Parties to the Treaty to develop research, production and use of nuclear energy

for peaceful purposes without discrimination and in conformity with Articles I and II of this Treaty.

2. All the Parties to the Treaty undertake to facilitate, and have the right to participate in the fullest possible exchange of equipment, materials and scientific and technological information for the peaceful uses of nuclear energy. Parties to the Treaty in a position to do so shall also cooperate in contributing alone or together with other States or international organizations to the further development of the applications of nuclear energy for peaceful purposes, especially in the territories of non-nuclear-weapon States Party to the Treaty, with due consideration for the needs of the developing areas of the world.

ARTICLE V

Each Party to the Treaty undertakes to take appropriate measures to ensure that, in accordance with this Treaty, under appropriate international observation and through appropriate international procedures, potential benefits from any peaceful applications of nuclear explosions will be made available to non-nuclear-weapon States Party to the Treaty on a non-discriminatory basis and that the charge to such Parties for the explosive devices used will be as low as possible and exclude any charge for research and development. Non-nuclear weapon States Party to the Treaty shall be able to obtain such benefits, pursuant to a special international agreement or agreements, through an appropriate international body with adequate representation of non-nuclear-weapon States. Negotiations on this subject shall commence as soon as possible after the Treaty enters into force. Non-nuclear-weapon States Party to the Treaty so desiring may also obtain such benefits pursuant to bilateral agreements.

ARTICLE VI

Each of the Parties to the Treaty undertakes to pursue negotiations in good faith on effective measures relating to cessation of the nuclear arms race at an early date and to nuclear disarmament, and on a treaty on general and complete disarmament under strict and effective international control.

ARTICLE VII

Nothing in this Treaty affects the right of any group of States to conclude regional treaties in order to assure the total absence of nuclear weapons in their respective territories.

ARTICLE VIII

1. Any Party to the Treaty may propose amendments to this Treaty. The text of any proposed amendment shall be submitted to the Depositary Governments which shall circulate it to all Parties to the Treaty. Thereupon, if requested to do so by one-third or more of the Parties to the Treaty, the Depositary Governments shall convene a conference, to which they shall invite all the Parties to the Treaty, to consider such an amendment.

2. Any amendment to this Treaty must be approved by a majority of the votes of all the Parties to the Treaty, including the votes of all nuclear-weapon States Party to the Treaty and all other Parties which, on the date the amendment is circulated, are members of the Board of Governors of the International Atomic Energy Agency. The amendment shall enter into force for each Party that deposits its instrument of ratification of the amendment upon the deposit of such instruments of ratification by a majority of all the Parties, including the instruments of ratification of all nuclear-weapon States Party to the Treaty and all other Parties which, on the date the amendment is circulated, are members of the Board of Governors of the International Atomic Energy Agency. Thereafter, it shall enter into force for any other Party upon the deposit of its instrument of ratification of the amendment.

3. Five years after the entry into force of this Treaty, a conference of Parties to the Treaty shall be held in Geneva, Switzerland, in order to review the operation of this Treaty with a view to assuring that the purposes of the Preamble and the provisions of the Treaty are being realised. At intervals of five years thereafter. a majority of the Parties to the Treaty may obtain, by submitting a proposal to this effect to the Depositary Governments, the convening of further conferences with the same objective of reviewing the operation of the Treaty.

ARTICLE IX

1. This Treaty shall be open to all States for signature. Any State which does not sign the Treaty before its entry into force in accordance with paragraph 3 of this Article may accede to it at any time.

2. This Treaty shall be subject to ratification by signatory States. Instruments of ratification and instruments of accession shall be deposited with the Governments of the United Kingdom of Great Britain and Northern Ireland, the Union of Soviet Socialist Republics and the United States of America, which are hereby designated the Depositary Governments.

3. This Treaty shall enter into force after its ratification by the States, the Governments of which are designated Depositaries of the Treaty, and forty other States signatory to this Treaty and the deposit of their instruments of ratification. For the purposes

of this Treaty, a nuclear weapon State is one which has manufactured and exploded a nuclear weapon or other nuclear explosive device prior to January 1, 1967.

4. For States whose instruments of ratification or accession are deposited subsequent to the entry into force of this Treaty, it shall enter into force on the date of the deposit of their instruments of ratification or accession.

5. The Depositary Governments shall promptly inform all signatory and acceding States of the date of each signature, the date of deposit of each instrument of ratification or of accession, the date of the entry into force of this Treaty, and the date of receipt of any requests for convening a conference or other notices.

6. This Treaty shall be registered by the Depositary Governments pursuant to Article 102 of the Charter of the United Nations.

ARTICLE X

1. Each Party shall in exercising its national sovereignty have the right to withdraw from the Treaty if it decides that extraordinary events, related to the subject matter of this Treaty, have jeopardized the supreme interests of its country. It shall give notice of such withdrawal to all other Parties to the Treaty and to the United Nations Security Council three months in advance. Such notice shall include a statement of the extraordinary events it regards as having jeopardized its supreme interests.

2. Twenty-five years after the entry into force of the Treaty, a conference shall be convened to decide whether the Treaty shall continue in force indefinitely, or shall be extended for an additional fixed period or periods. This decision shall be taken by a majority of the Parties to the Treaty.

ARTICLE XI

This Treaty, the English, Russian, French, Spanish and Chinese texts of which are equally authentic, shall be deposited in the archives of the Depositary Governments. Duly certified copies of this Treaty shall be transmitted by the Depositary Governments to the Governments of the signatory and acceding States.

IN WITNESS WHEREOF the undersigned, duly authorised, have signed this Treaty.

DONE in triplicate, at the cities of London, Moscow and Washington, the first day of July, one thousand nine hundred and sixty-eight.

Index

Note: page numbers in *italic type* refer to Tables; those followed by "n" refer to notes at the bottom of that page.

Acheson, Dean 97
ADA *see* Atomic Development Authority
Advanced Fuel Cycle Initiative (US) 139–40
Afghanistan 22, 94, 124, 144
Ahmadinejad, President Mahmoud 110
air burst nuclear explosions 171, 172, 175
al-Qaeda 20, 22, 25, 36, 38, 94, 116, 124, 144
Argentina 37, 122
aspirant nuclear states 21, 26–7, 37, 87, 102, 121, 122, 159, 161
atmospheric pollution 139
atmospheric radiation levels 75, 138, 139
Atomic Development Authority (ADA) 81–2
"Atoms for Peace" program 82–3, 90, 91, 104, 119, 142

Ballistic Missile Defense system (US) 14, 98, 104, 107, 110, 111, 127, 166
Ban Ki-moon 64
Bangla Desh 65
Baruch Plan 81–2
BBC China 67, 120
Belarus, research reactors 26
Belfer Centre for Science and International Affairs 126
Belgium:
- export of nuclear technology 120
- research reactors 24
- stocks of fissile material 28

Bhutto, Zulfikar Ali 65
biological weapons *see* weapons of mass destruction
blast (nuclear explosions) 11, 170, 171, 173, 174
Bodman, S. 139
Bratislava Summit on nuclear security (Russia/US, 2005) 142
Brazil 3, 37, 122, 158
BRIC states (Brazil, Russia, India, China) 3
Brown, Prime Minister Gordon 179
Buffet, Warren 136
Bull, Hedley 101
Bunn, Matthew 51, 62
burn injuries from nuclear explosions 174
Bush, President George W. 91, 142, 176
- Bush Administration 14, 19, 55, 131, 139, 166

Buzan, Barry 99, 100

Canada 158
- CANDU (Canadian Deuterium-Uranium) technology 7, 89–90
- Manhattan Project 79
- membership of UNAEC 81
- nuclear power generation 36, 69
- research reactors 24
- supply of nuclear materials to India 37, 89–90, 91
- use of HEU in production of medical applications 32

centrifuges 7, 22, 148
- illicit trade in 36, 37, 55, 60, 65–7, 77, 97, 102, 120, 121
- possession by Iran 66, 104, 105, 106, 107, 110, 178

chain reactions 5, 9–10, 34, 60, 175
chemical weapons *see* weapons of mass destruction
Chernobyl nuclear accident 170
China 3, 93

alliance with North Korea 85
and NPT 24
nuclear support for Pakistan 94
nuclear warheads *31*
plutonium reactors 87
plutonium stocks 30
Six Party nuclear negotiations 26
stocks of fissile material 28
war with India 88
CIRUS research reactor, India 90, 91
Civil Support (US Homeland Security Act) 2, 19
clandestine nuclear states 21, 26–7, 102, 122, 157
climate change 149
and nuclear energy 129, 161–2
Code of Conduct for the Safety and Security of Radioactive Sources (IAEA, 2004) 45
Cold War 82, 97, 99, 138, 154, 163
end of 3, 49, 62, 100, 158
Commission on the Prevention of Weapons of Mass Destruction Proliferation and Terrorism 94
complacency regarding nuclear risks 1, 20, 64, 156
Comprehensive Test Ban Treaty (1996) 131–2, 152, 177
Congressional Office of Technology Assessment (US) 124
Convention on the Law of the Sea (LOSC, 1982) 42, 46, 55
Convention on Nuclear Safety (1994) 45, 152
Convention on the Physical Protection of Nuclear Materials (2005) 54, 178
Convention on the Physical Protection of Nuclear Materials and Nuclear Facilities (1979) 45
Convention for the Suppression of Acts of Nuclear Terrorism 178
Cooperative Threat Reduction Program (CTR, Russia/US) 144
Copenhagen School 100
counter-terrorism 115, 116–17, 126–8
evaluation of strategies 128–32
Counter-Terrorism and Strategic Security Working Group (CTWG, Russia/US) 103–4, 106, 111
CTR *see* Cooperative Threat Reduction Program
CTWG *see* Counter-Terrorism and Strategic Security Working Group
Cuban Missile Crisis (1962) 84, 109
Czech Republic, and US missile defense shield 14, 98, 99, 110, 166

Department of Defense (DOD, US) 2, 19, 35
Department of Homeland Security (DHS, US) 128, 146, 168
Design Basis Threat assessment 63
detonators 110–11
deuterium 79n1
see also CANDU technology
developing states with nuclear weapons 25–6
DHS *see* Department of Homeland Security
dirty bombs 130, 142
disarmament *see* nuclear disarmament
DOD *see* Department of Defense
down-blending of HEU 14–15, 29, 56, 77–8, 87, 98, 134, 144, 154
DPRK *see* North Korea
Dubai 66, 67

Eastern Europe 146
NATO expansion into 14, 98
US missile defense system 14, 98, 104, 107, 110, 111, 127, 166
Egypt 37, 110, 122
Eighteen Nations Disarmament Committee 138
Eisenhower, President D. 82–3, 91, 104
ElBaradei, Mohamed 71, 72, 84, 135, 157
nuclear fuel bank proposals 136–7, 140, 161–2
electromagnetic pulse (EMP) 11, 170, 173
energy infrastructure 101
established nuclear states 24
European Union, contribution to NTI fuel bank initiative 137

export controls on nuclear materials 16, 44, 53, 55, 86, 105, 106, 108, 121, 155

fallout *see* nuclear fallout
fast breeder reactors 30, 72, 73, 160
fast neutron breeder reactors 89
"Fat Man" nuclear bomb (Nagasaki) 79
Federally-Administered Tribal Areas (FATA), Pakistan 94
Finland, nuclear waste disposal 8
fireballs (nuclear explosions) 11, 171, 173, 175–6
First Gulf War 122
First World War 41
fissile material:
  access to 23, 116, 117, 124, 136, 137, 140, 150
  covert trade in 36–7
  Fissile Material Cut-Off Treaty (FMCT) 71, 92, 115, 125, 130, 134, 136, 152, 158, 177
  Global Fissile Material Report (2007) 6
  IAEA monitoring 33, 156–7
  International Panel on Fissile Materials 6, 29, 33, 70, 160
  laws relating to security of 45
  state-based diversion of 67–9
  stocks of 6, 61, 141–4, 154
  summary of current situation 27–32
  weapons grade, security and monitoring 32–3
  see also nuclear fuel; plutonium; uranium
flash (nuclear explosions) 11, 169, 174
food security 115
France:
  cooperation agreement with India 25
  export of nuclear technology 120
  and NPT 24
  nuclear power generation 36, 69
  nuclear reactors 145
  nuclear reprocessing 32, 73, 160
  nuclear support for Iran 104
  nuclear warheads *31*
  plutonium stocks 24, 30
  stocks of fissile material 28
  use of MOX fuel 32
fuel bank proposals 15, 70–2, 82, 85, 91, 131, 132, 133–41, 152, 154, 155, 161–2
  NTI International Fuel Bank initiative 16, 46–7, 136–7, 138, 139, 140–1, 147, 153
  Obama's proposal 115, 125
fuel rods 7, 53
  spent 8, 23, 32, 63, 72–3
  see also nuclear reprocessing; spent fuel

G8, Global Partnership Against the Spread of Weapons and Materials of Mass Destruction 13, 52, 56–7, 125
Gates, Robert 118, 166
General Assembly *see* United Nations General Assembly
General Welfare (provision of US Homeland Security Act) 2, 19
Generation IV reactors 78, 95, 145–6, 154, 155
Geneva Conventions 42
geological sequestration 8, 139, 145, 160
Georgia, nuclear security 62, 63
Germany 158
  export of nuclear technology 53, 120
  fuel bank proposals 135–6
  nuclear power generation 36, 69
  nuclear support for Iran 104
  plutonium stocks 30
  research reactors 24
  stocks of fissile material 28
Ghandi, Rajiv 88
Global Fissile Material Report (2007) 6
Global Initiative to Combat Nuclear Terrorism (Russia/US) 13, 52, 53–4, 115, 125, 142–3
Global Nuclear Energy Partnership (GNEP) 15, 47, 139
Global Partnership Against the Spread of Weapons and Materials of Mass Destruction (G8) 13, 52, 56–7, 125
Global Summit on Nuclear Security (2010) 116, 125, 177
Global Threat Reduction Initiative (Russia/US, 2004) 142
Global Threat Reduction Program (US) 30

global warming *see* climate change
globalization 3
GNEP *see* Global Nuclear Energy Partnership
Greece 41
  research reactors 30
ground shock (nuclear explosions) 11, 170, 172, 173

H bombs 10, *31*, 79
Hague Code of Conduct Against Ballistic Missile Proliferation (HCOC) 105, 107
heavy water reactors 110
high altitude burst nuclear explosions 173
Hiroshima bombing 10, 34, 79, 170, 174
Hizballah 20
Homeland Defense (US Homeland Security Act) 2, 19
homeland security 12, 54, 100
  United States 1–3, 19–21, 100, 126, 128, 129, 144, 168
Homeland Security Act 2002 (US) 2, 19, 74
Homeland Security Institute (US) 2
Hydrogen bombs 10, *31*, 79

IAEA (International Atomic Energy Agency) 30, 44–5, 75, 83–4, 101, 130, 134–5
  Additional Protocols 157, 178
  Iran 104, 109
  Code of Conduct for the Safety and Security of Radioactive Sources (2004) 45
  cooperation agreement with India 25
  and fuel bank proposals 15, 16, 46–7, 84–5, 140–1
  Illicit Trafficking Data Base (ITDB) 63
  inspection and verification role 27, 49, 64, 74, 91, 115, 125
  India 92
  Iran 105–6, 106–7, 156–7, 158–9
  and Iraq 97–8
  and Libya 102–3
  monitoring of fissile material stocks 33, 156
  and nuclear security 15, 62–3, 143
  resourcing of 178
  Safeguards Agreements 44, 69, 104, 152
Illicit Trafficking Data Base (IAEA) 63
improvised nuclear devices (INDs) 60, 117, 123, 124, 156
India 3, 9, 37, 49, 53, 55, 65, 80, 122, 141, 158
  CIRUS research reactor 90, 91
  criticism of NPT 48–9, 112
  IAEA inspection 92
  marine nuclear propulsion 34
  non-membership of NPT 43n3, 89, 91, 93
  nuclear agreement with US (2008) 15, 90, 91–3, 95
  nuclear energy program 89
  and nuclear fuel cycle 88–93
  nuclear security 25, 35
  nuclear warheads *31*
  nuclear weapons testing 9, 25, 37, 65, 84, 88, 90, 921
  opposition to fuel bank proposals 140
  plutonium reactors 87, 89
  plutonium stocks 30
  production of fissile materials 33–4
  relationship with Nuclear Suppliers Group (NSG) 25, 92, 93
  stocks of fissile material 28
  use of CANDU (Canadian Deuterium-Uranium) technology 89–90
  use of thorium 77, 88, 89
  war with China 88
Indonesia 55
  nuclear power generation 70
INDs *see* improvised nuclear devices
"insiders" as a threat to nuclear security 23, 35–6, 62, 118
intelligence gathering 163
International Atomic Energy Agency *see* IAEA
International Convention for the Suppression of Acts of Nuclear Terrorism (2005) 45, 54
International ("Hague") Code of Conduct Against Ballistic Missile Proliferation (HCOC) 105, 107

International Institute for Strategic Studies 121, 122
international law 13, 41–3, 151–3
  and nuclear fuel security 46–7, 49–50
  nuclear law 43–6
International Nuclear Fuel Authority 91
International Panel on Fissile Materials (IPFM) 6, 29, 33, 70, 160
"International society" (Bull) 101
International Uranium Enrichment Center 135
ionizing radiation 11, 170, 171, 173, 174
IPFM *see* International Panel on Fissile Materials
Iran 102, 131, 166
  IAEA inspections 104, 109
  import of nuclear materials and technology 53, 55, 60, 66, 121, 122
  international policy towards 52, 126, 127, 128–9
  missile capability 98, 104–5, 107–10
  and NPT 104, 105, 109
  nuclear activities 14, 26, 103–12
  nuclear power generation 70
  space program 107, 108
  uranium enrichment operations 95, 98, 104, 105, 106, 110, 178–9
Iraq:
  2003 invasion of 163
  development of nuclear capacity 67, 97–8, 121, 122–3
Israel 9, 26–7, 37, 49, 98, 110, 122, 141
  non-membership of NPT 43n3
  nuclear warheads *31*
  plutonium stocks 30
  stocks of fissile material 28
Italy, research reactors 30
ITDB *see* Illicit Trafficking Data Base

Japan 158
  Generation III and 3+ reactors 145
  HEU enrichment capacity 70–2
  Hiroshima bombing 10, 34, 79, 170, 174
  Nagasaki bombing 79, 170, 174
  nuclear power generation 3, 36, 69
  nuclear reprocessing 32–3, 73, 160
  plutonium stocks 30, 33
  research reactors 24
  Six Party nuclear negotiations 26
  stocks of fissile material 28
  use of MOX fuel 32–3
Joint Convention on the Safety of Spent Fuel Management and on the Safety of Radioactive Waste Management (1997) 45
Joseph, Robert G. 166

Kahn, Herman 133
Kennedy, President John F. 49, 50, 75, 109, 112
Khan, A.Q., nuclear trafficking network 6, 13, 15, 25, 36, 60, 65–7, 93–4, 102, 104, 113, 119, 120–2, 155
Kim Il-sung 86
Korean War 85
Kuwait, contribution to NTI fuel bank initiative 137

law, international *see* international law
Law of Nations *see* international law
Law of the Sea Convention (LOSC, 1982) 42, 46, 55
Lebanon 20
Lewis, Jeffrey 123–4
Libya 37, 55, 60, 66–7, 77, 102–3, 120, 122
light water reactors 53, 70, 77, 139
"Little Boy" nuclear bomb (Hiroshima) 79
London bombings (2005) 116
London Suppliers Group *see* Nuclear Suppliers Group
LOSC *see* Law of the Sea Convention
Lugar, Richard 144

McNamara, Robert 109
Madrid bombings (2004) 116
Malaysia 55, 67, 77
Manhattan Project 79
marine pollution 44
maritime law, Law of the Sea Convention (LOSC, 1982) 42, 46, 55
Matrix Churchill 98, 122
medical uses of HEU 32, 40, 119, 131, 141

Megaports initiative (US) 146
misperception, dangers of 95, 126, 129
missile defense system (US) 14, 98, 104, 107, 110, 111, 127, 166
Missile Technology Control Regime (MTCR) 105, 107, 108
MNAs *see* multilateral nuclear approaches
molybdenum-99 141
MOX (mixed oxide) fuel 8–9, 32–3, 63, 70, 72–3, 139, 160
MTCR *see* Missile Technology Control Regime
multilateral nuclear approaches (MNAs) 84–5, 91, 95
mushroom clouds 11, 171
mutually assured destruction 100

Nagasaki bombing 79, 170, 174
National Nuclear Security Agency (NNSA, US) 25, 30, 35
National Research Council (US) 48
national security 21, 73–4, 100
National Security Advisory Group (US) 126
National Security (provision of US Homeland Security Act) 2, 19
National Strategy for Homeland Security (US) 19–21
NATO (North Atlantic Treaty Organization) 24
  expansion into Eastern Europe 14, 98
naval vessels, nuclear propulsion of 29–30, 32, 34, 56, 141
Netherlands 65
  fuel bank proposals 135–6
  research reactors 24
NNSA *see* National Nuclear Security Agency
Non-Nuclear Weapon States (NNWS) 44, 68, 102, 112, 158
non-state nuclear actors 27
Nonproliferation Policy Education Center 92
North Korea 9, 14, 26, 37, 49, 55, 60, 66, 80, 102, 122, 131, 141, 144, 166
  alliance with China 85
  and IAEA 86
  international policy towards 52, 126, 127, 128–9
  and NPT 43n3, 86, 88
  and nuclear fuel cycle 85–8
  nuclear support from Soviet Union 85–6
  nuclear warheads *31*
  stocks of fissile material 28
  supply of missile technology to Iran 107–8
Norway, contribution to NTI fuel bank initiative 137
NPT (Nuclear Non-proliferation Treaty) 24, 25, 33, 36, 38, 43–6, 48, 68–9, 74–5, 83, 102, 135, 137–8, 152
  2010 Review Conference 64, 155, 178
  Article III 69, 158, 183
  Article IV 136, 140, 183–4
  assessment of effectiveness of 51–3
  criticisms of 48–9, 112
  enforcement 115, 125, 130, 138
  and Iraq 122
  and Libya 102
  and North Korea 26
  and nuclear disarmament 43, 44, 48, 49, 102
  and Pakistan 43n3, 94
  strengthening of 177–8
  text of 181–6
NSG *see* Nuclear Suppliers Group
NTI *see* Nuclear Threat Initiative
nuclear disarmament 52, 71, 82, 138, 141, 155, 177
  and the NPT 43, 44, 48, 49, 102
nuclear energy *see* nuclear power generation
nuclear explosions, effects of 10–12, 169–73
  delayed effects 175–6
  immediate effects 173–5
nuclear fallout 11, 170, 171–2, 173, 175–6
nuclear fission 5, 9–10
Nuclear Free Zones 44
nuclear fuel *see also* fissile material; plutonium; uranium
nuclear fuel cycle 6–9, 69–73
  Advanced Fuel Cycle Initiative (US) 139–40

and India 88–93
and North Korea 85–8
and Pakistan 93–5
and proliferation 87–8
nuclear fuel security 12, 13, 21–2, 39–41, 59–60, 99, 148
categories of threat 31–8, 61–78
historical record 14, 79–85, 79–95
illegal trafficking 63–7
multilateral nuclear approaches (MNAs) 84–5, 91, 95
policy and practice 73–5
reappraisal of 156–9
scope of the issues 22–4
state-based diversion of fissile materials 67–9
and strategic security 102, 113
strategies for 133–48, 165–80
theft 61–7
threats 13–14
triage approach to threats 76–8
see also nuclear security; strategic security
Nuclear Fuel Supply: Challenges and Opportunities 135–6
nuclear fusion 10
nuclear non-proliferation 13
and nuclear fuel cycle 87–8
proliferation-resistant technologies 15, 16, 78, 87, 95, 128, 131, 144, 160, 162
strategies 15, 16, 133–48
see also NPT (Nuclear Non-Proliferation Treaty)
Nuclear Non-proliferation Act (US, 1978) 90–1
nuclear power generation 5, 16, 36, 69–70
and climate change 161–2
expansion of 127, 129, 139, 159–61
nuclear propulsion of naval vessels 29–30, 32, 34
nuclear reactors 144–6, 158
fast breeder 30, 72, 73, 160
fast neutron breeder 89
Generation III and 3+ 145
Generation IV 78, 95, 145–6, 154, 155
heavy water 110
light water 53, 70, 77, 139
for power generation 159
proliferation-resistant technologies 78, 87, 95, 128, 131, 144–5, 145, 154
Teller-Ulam design 10
see also research reactors
Nuclear Regulatory Commission (US) 139
nuclear reprocessing 8, 24, 29, 32, 47, 72–3, 87, 91, 139, 158–9
see also nuclear waste; spent fuel
nuclear security 99, 118
categories of threat 31–8
enforcement measures 45
future policies 149–63
international policy framework 47–50
success of 151–6
see also nuclear fuel security; strategic security
nuclear smuggling 23, 51, 78, 146–7, 154, 155
see also Khan, A.Q. nuclear trafficking network
Nuclear Suppliers Group (NSG) 27, 37, 44, 89, 90, 121, 155
relationship with India 25, 92, 93
Trigger Lists 44, 90, 121, 141, 155
nuclear terrorism 15, 20, 21, 22–3, 31–2, 48, 51, 54, 94–5, 113, 115–17, 143
access to nuclear weapons 35–6, 51, 117–20, 128
assessment of threat 123–5
counter-terrorism 115, 116–17, 126–32
Global Initiative to Combat Nuclear Terrorism 13
home-grown 20
international suppression of 44
recruitment of skilled personnel 34–5, 51, 52
nuclear testing 44, 50, 75, 138, 177
Comprehensive Test Ban Treaty (1996) 131–2, 152, 177
Nuclear Test Ban Treaty 75
Nuclear Threat Initiative (NTI) 51
International Fuel Bank initiative 16, 46–7, 136–7, 138, 139, 140–1, 147, 153
nuclear waste 8, 72–3, 139, 145, 148, 151
Joint Convention on the Safety of Spent Fuel Management and on

the Safety of Radioactive Waste Management (1997) 45
see also nuclear reprocessing; spent fuel
Nuclear Weapon States (NWS) 44, 49, 102, 112, 141, 153, 158, 158–9
nuclear weapons:
acquisition by terrorists 35–6, 51, 117–20, 128
definition 79n1
development by states 37–8
dirty bombs 130, 142
Hydrogen bombs 10, *31*, 79
improvised nuclear devices 60, 117, 123, 124, 156
number of warheads *31*
technologies 9–12
testing 44, 50, 75, 131–2, 138, 152, 177
thermonuclear weapons 10, 79
Nuclear Weapons Convention 131
nuclear-interested states:
aspirant and clandestine nuclear states 26–7
developing states with nuclear weapons 25–6
established nuclear states 24
non-state nuclear actors 27
Nunn, Sam 144
NWS *see* Nuclear Weapon States

Obama, President Barack:
nuclear policy 115–16, 125, 126, 131–2, 165–6, 177, 179
Obama Administration 16, 21, 73, 131–2, 176
Omid satellite (Iran) 108
Osama bin Laden 120, 123, 124

Pakistan 9, 14, 37, 49, 53, 80, 141, 144
and al-Qaeda 22, 25, 36, 38
A.Q. Khan nuclear trafficking network 6, 13, 15, 25, 36, 60, 65–7, 93–4, 102, 104, 113, 119, 120–2, 155
non-membership of NPT 43n3, 94
and nuclear fuel cycle 93–5
nuclear power generation 70
nuclear security 22, 25, 35–6, 38
nuclear support from China 94
and nuclear terrorism 94–5
nuclear warheads *31*
plutonium stocks 30–1
production of fissile materials 33–4
stocks of fissile material 28, 30–1
Peaceful Nuclear Explosion (PNE) *see* India, nuclear weapons testing
Pelindaba, South Africa, attack on nuclear facility 62–3, 119, 142
plutonium 7–8, 160–1
disposal of excess by Russia and US 76–8
global stocks 28–9, *28*, 141
France 24
Russia 30, 33
separated plutonium 30–1
UK 24, 33
illicit trafficking of 63–7
improvised nuclear devices 60
low-enriched 23
nuclear security risks 60, 81
number of warheads *31*
plutonium-239 isotope 8, 22, 23, 24, 79
reactors 87, 89
reprocessing 72–3, 133–4, 139, 160
roles and uses 5–6
threat posed by continuing production 33–4
use in MOX (mixed oxide) fuel 8–9, 32, 63, 72–3, 139, 160
use in nuclear power generation 22
see also fissile material; nuclear fuel; uranium
Poland, and US missile defense shield 14, 98, 110, 166
pollution, marine 44
Portugal, research reactors 30
power parks 85
proliferation *see* NPT; nuclear non-proliferation
Proliferation Security Initiative (PSI, US) 13, 27, 37, 45–6, 52, 55–6, 78, 108, 115, 125, 131, 146, 155, 166
proliferation-resistant technologies 15, 16, 78, 87, 95, 128, 131, 144, 160, 162
PSI *see* Proliferation Security Initiative

Putin, President Vladimir 142

radiation sickness 174
radioactivity 7
  of nuclear waste 8, 28, 32, 72, 73, 130, 145, 148, 160
  radioactive fallout 11, 170, 171–2, 173, 175–6
reactors *see* nuclear reactors
research reactors 6, 22, 24, 26, 32, 56, 63–4, 87, 141–2
  conversion to LEU 134, 142, 143
  reduction of HEU in 30
  see also nuclear reactors
  Russia 3, 54
  Bratislava Summit on nuclear security (2005) 142
  Cooperative Threat Reduction Program 144
  Counter-Terrorism and Strategic Security Working Group (CTWG) 103–4, 106, 111
  disposal of excess plutonium 76
  down-blending of HEU 14–15, 29, 56, 77–8, 98, 144, 154
  fuel bank proposals 135
  Global Initiative to Combat Nuclear Terrorism 13, 52, 53–4, 115, 125, 142–3
  Global Threat Reduction Initiative (2004) 142
  and NPT 24
  nuclear cooperation with US 142–3, 144
  nuclear disarmament 71, 144, 154, 155, 177
  nuclear policy 98
  nuclear power generation 36, 69
  nuclear reprocessing 73, 160
  nuclear security 3, 62, 63, 118, 144, 146
  nuclear support for Iran 104
  nuclear warheads *31*
  plutonium stocks 30, 33, 56
  production of fissile material 34
  research reactors 30
  Six Party nuclear negotiations 26
  stocks of fissile material *28*, 29
  for use in naval vessels 29–30, 32
  Strategic Offensive Reduction Treaty (2003) 31, 155
  see also Soviet Union

Saddam Hussein 97, 122
Safeguards Agreements (IAEA) 44, 69, 104, 152
Safir rocket (Iran) 107, 108
Sarkozy, President Nicolas 179
Saudi Arabia 110
Second Line of Defense initiative (US) 146
Second World War 42, 79, 81, 156, 174
"Securing the Bomb 2008" (NTI Report) 51
Security Council *see* United Nations Security Council
sequestration of nuclear waste 8, 37, 145, 148, 154
Shahab 3M missile (Iran) 108
shock (nuclear explosions) 11, 173
Singh, Prime Minister Manmohan 91
Six Party nuclear negotiations 26
Sokolski, Henry 92
SORT *see* Strategic Offensive Reduction Treaty
South Africa 37, 102, 122
  attack on Pelindaba nuclear facility 62–3, 119, 142
  research reactors 26, 119
South Korea 85, 86
  nuclear power generation 36, 69
  research reactors 30
  Six Party nuclear negotiations 26
Soviet Union:
  and "Atoms for Peace" program 82
  Cuban Missile Crisis (1962) 84, 109
  and formation of UNAEC 81
  nuclear disarmament 138
  nuclear security in former states of 87, 142, 146
  nuclear support for North Korea 85
  nuclear weapons 10
  rejection of Baruch Plan 82
  see also Russia
Spain, research reactors 30
spent fuel 8, 23, 32, 72–3, 91, 133–4, 139, 160

Joint Convention on the Safety of Spent Fuel Management and on the Safety of Radioactive Waste Management (1997) 45
see also nuclear reprocessing; nuclear waste
START 1 Treaty 155
Stern Review of the Economics of Climate Change (UK) 162
stockpiles of nuclear material 6, 27–32, 52, 130
Russia 22, *28*
US 15, *28*
Strategic Offensive Reduction Treaty (Russia/US, 2003) 31, 155
strategic security:
and enriched uranium 14–15, 97–113
global threats and opportunities 165–80
and nuclear fuel security 102–3
see also nuclear fuel security; nuclear security
sub-surface burst nuclear explosions 172–3
surface burst nuclear explosions 172
Sweden
nuclear waste disposal 8
research reactors 30
Switzerland, stocks of fissile material 28
Syria 37, 122

Taliban 25, 94, 124–5, 144
Teller-Ulam reactor design 10
terrorism *see* nuclear terrorism
testing *see* nuclear testing
thermal pulse 170
thermal radiation 11, 171, 173, 174
thermonuclear weapons 10, 79
thorium 63, 77, 88, 89
TNT (trinitrotoluene) explosive 10
Track 2 threat assessment 103, 106
trafficking in nuclear fuels 63–7
transportation:
of nuclear material and technology 44, 45–6, 146, 155
of uranium 7
triage approach to threat response 76–8
Trigger Lists 44, 90, 121, 141, 155
tritium 79n1
"trust but verify" policy 16, 69, 153

Ukraine 56
nuclear power generation 36, 69
nuclear security 62
research reactors 26
United Arab Emirates, contribution to NTI fuel bank initiative 137
United Kingdom:
export of nuclear technology 120
to Iraq 97–8
first generation reactors 145
and formation of UNAEC 81
fuel bank proposals 135–6
Manhattan Project 79
and NPT 24
nuclear reprocessing 32, 73, 160
nuclear warheads *31*
nuclear-powered submarines 29–30, 32
plutonium stocks 30, 33
Stern Review of the Economics of Climate Change 162
stocks of fissile material 28, 29–30
plutonium stocks 24
terrorist attacks 116–17
United Nations 64, 101, 128, 134
Charter 42, 146
General Assembly 81, 82–3, 134, 165, 177
Nuclear Weapons Convention 154
Security Council 81, 82, 83, 151
resolution 1540 (2004) 37, 45, 54, 88, 116, 130, 146, 154, 178
resolution 1690 (2006) 105
resolution 1887 (2009) 165, 177–9
United Nations Atomic Energy Commission (UNAEC) 81, 134
UNMOVIC 122
United States 54, 55
Advanced Fuel Cycle Initiative 139–40
Bratislava Summit on nuclear security (2005) 142
Congressional Office of Technology Assessment 124
Constitution 2, 19

Cooperative Threat Reduction Program 144
counter-terrorism 115
Counter-Terrorism and Strategic Security Working Group (CTWG) 103–4, 106, 111
Cuban Missile Crisis (1962) 84, 109
Department of Defense 2, 19, 35
Department of Homeland Security 128, 146, 168
disposal of excess plutonium 76
down-blending of HEU 14–15, 29, 56, 77–8, 87, 98, 144, 154
and formation of UNAEC 81
Global Initiative to Combat Nuclear Terrorism 13, 52, 53–4, 115, 125, 142–3
Global Threat Reduction Initiative (2004) 142
Global Threat Reduction Program 30
Homeland Security 1–3, 100, 126, 128, 129, 144, 168
National Strategy for Homeland Security 19–21
Manhattan Project 79
Megaports initiative 146
missile defense system 14, 98, 104, 107, 110, 111, 127, 166
National Nuclear Security Agency 25, 30, 35
National Research Council 48
National Security Advisory Group 126
and NPT 24
and NTI fuel bank initiative 137, 139, 140
nuclear agreement with India (2008) 15, 90, 91–3, 95
nuclear cooperation with Russia 142–3, 144
nuclear disarmament 71, 82, 134, 138, 155, 177
Nuclear Non-proliferation Act (1978) 90–1
nuclear policy 80, 93, 97, 139–40, 166–9, 177–9
nuclear power generation 36, 69, 159
nuclear reactor technology 145
Nuclear Regulatory Commission 139
nuclear reprocessing 14–15, 29, 56, 73, 77–8, 87, 98, 144, 154
nuclear security 22, 118, 125–6, 128–32, 167–9
nuclear smuggling 146
nuclear stockpile 15
nuclear support for Iran 104
nuclear warheads *31*
nuclear waste disposal 8
Proliferation Security Initiative 13, 27, 37, 45–6, 52, 55–6, 78, 108, 115, 125, 131, 146, 155, 166
Second Line of Defense initiative 146
Six Party nuclear negotiations 26
stocks of fissile material *28*
Strategic Offensive Reduction Treaty (2003) 31, 155
uranium 7
HEU (highly enriched uranium) 6
enrichment capacity 70–2, 160
global stocks 28, *28*, 29–30, 141
illicit trafficking of 63–7
improvised nuclear devices 60
medical applications of 32
nuclear security risks 59–60, 70–2, 81
number of warheads *31*
roles and uses 5–6
and strategic security 14–15, 97–113
threat posed by continuing production 33–4
use for research 63–4
LEU (low enriched uranium) 23, 30, 63, 70, 77, 105, 106, 110, 134, 135, 142, 143, 148, 155, 160
MOX (mixed oxide) fuel 8–9, 32–3, 63, 70, 72–3, 139, 160
U-233 isotope 77
U-235 isotope 5, 7, 8–9, 10, 23, 37, 72, 79, 106
U-238 isotope 7, 72
uranium hexafluoride 7, 22, 62, 105, 107
yellowcake (uranium oxide concentrate) 7, 92
see also fissile material; nuclear fuel; plutonium

Vanunu, Mordechai 27
Vienna Convention on Civil Liability for Nuclear Damage (1977) 45

Washington Summit Meeting on Nuclear Security (2010) 116, 125, 177
Wassenaar Arrangement 44, 121, 155
waste *see* nuclear waste
water shock (nuclear explosions) 172
weapons of mass destruction 20, 94
Westphalia, Peace of 41
Westphalian states 3, 17, 23
Working Group on Counter-Terrorism and Strategic Security (CTWG, Russia/US) 103–4, 106, 110
World Nuclear Association (WNA) 7, 159

yellowcake (uranium oxide concentrate) 7, 92

Zangger Committee 44, 121, 155
Zimmerman, Peter 123–4